Farmers and Scientific Fish Culture

The Authors

Dr. Biswajit Goswami is currently employed as Subject Matter Specialist in Fisheries Science of Dakshin Dinajpur Krishi Vigyan Kendra, Uttar Banga Krishi Viswavidyalaya, W.B., India. He did M.F.Sc. in Fishery Extension from WBUAFS, Kolkata and Ph.D. in Agricultural Extension from Visva Bharati University. He obtained his Post Graduate Diploma in Rural Development from IGNOU and Post Graduate Diploma in Agricultural Extension Management from MANAGE, Hyderabad. He was qualified NET examination in Fish and Fishery Sc. conducted by ASRB, ICAR. He has coordinated training and extension activities at grass root level in West Bengal. Dr.Goswami has specialisation in Training, Teaching, Research, Transfer of Technology, Communication and Management. He has authored 4 books (International) in fishery extension and contributed chapters in four books. He published twenty four research papers in Indian and foreign journals. He has been invited as an expert in TV and radio program. He is life Member of International Society For Research And Development (ISRD), UK , Inland Fisheries Society of India, CIFRI , Barrackpore, Crop And Weed Science Society (CWSS), BCKV, Society of Extension Education, Agra. He is recipient of Second best Poster presentation award in International Seminar on ISIAAR organized by Bidhan Chandra Krishi Viswavidyalaya. He is the fellow of INTERNATIONAL SOCIETY FOR RESEARCH AND DEVELOPMENT (ISRD), UK. The Search Committee constituted by Confederation of Indian Universities, New Delhi and World Achievers Foundation, Kolkata has selected his name for conferring upon him the Swami Vivekananda Excellence Award 2017.

Prof. (Dr.) Swapan Behari Mukhopadhyay is a retired Professor of Agricultural Extension from the Institute of Agriculture (Palli Siksha Bhavana) VISVA BHARATI. He is a person with long experience in Teaching, Research and Extension over more than forty years. He has also served North Bengal Campus, Bidhan Chandra Krish iViswavidyalaya now Uttar Banga Krishi Viswavidyalaya as the first Lecturer of Agril. Extension. As a Project Leader, some Central and State Govt. sponsored Evaluation and other Projects on Crop development were successfully conducted and reported by him. Professor Mukhopadhyay has guided six Ph.D. scholars and some MSc thesis works.

Prof. (Dr.) Shyam Sundar Dana is working as a Dean, Faculty of Fishery Sciences, West Bengal University of Animal and Fishery Sciences, Kolkata, West Bengal, India. He did his M.V.Sc and Ph.D in Extension Education from Indian Veterinary Research Institute, Izatnagar, Bareilly, Uttar Pradesh, India. He was awarded ICAR-in service Senior Research Fellowship and Best Student Award in Ph.D. He had undergone training at IIM, Ahmedabad, collaborative training programme organized by IARI, New Delhi and MSU, Michigan, USA. Prof. Dana worked in various capacities as Scientist at ICAR and Professor and Head, Department of Fishery Extension, Faculty of Fishery Sciences, West Bengal University of Animal and Fishery Sciences, Kolkata, West Bengal, India. He supervised 35 Master Degree students and 4 Ph. D students in Fishery Extension and published 36 research papers in Indian and Foreign journals and also published 8 books chapters in different books of Extension Education.

Farmers and Scientific Fish Culture

– Authors –

Dr. Biswajit Goswami
Subject Matter Specialist (Fisheries Science)
Dakshin Dinajpur Krishi Vigyan Kendra
Uttar Banga Krishi Viswavidyalaya

Prof. (Dr.) Swapan Behari Mukhopadhyay
Former Professor, Dept. of Agril. Extension,
Agril. Economics and Agril. Statistics (EES)
Palli Sikha Bhavan (Institute of Agriculture)
Visva Bharati University

Prof. (Dr.) Shyam Sundar Dana
Professor, Dept. of Fishery Extension
F/O Fishery Sciences
West Bengal University of Animal and Fishery sciences

2017

Scholars World

A Division of

Astral International Pvt. Ltd.

New Delhi – 110 002

Publisher's Note:

Every possible effort has been made to ensure that the information contained in this book is accurate at the time of going to press, and the publisher and author cannot accept responsibility for any errors or omissions, however caused. No responsibility for loss or damage occasioned to any person acting, or refraining from action, as a result of the material in this publication can be accepted by the editor, the publisher or the author. The Publisher is not associated with any product or vendor mentioned in the book. The contents of this work are intended to further general scientific research, understanding and discussion only. Readers should consult with a specialist where appropriate.

Every effort has been made to trace the owners of copyright material used in this book, if any. The author and the publisher will be grateful for any omission brought to their notice for acknowledgement in the future editions of the book.

Cataloging in Publication Data--DK
Courtesy: D.K. Agencies (P) Ltd. <docinfo@dkagencies.com>

Goswami, Biswajit, author.
Farmers and scientific fish culture / authors, Dr. Biswajit Goswami, Prof. (Dr.) Swapan Behari Mukhopadhyay, Prof. (Dr.) Shyam Sundar Dana.
pages cm
Includes bibliographical references.
ISBN 9789389605198 (hardback)

1. Fish culture--India--Dakshin Dinajpur (District) 2. Fish culture--India--Uttara Cabbiśa Paragaṇā. 3. Fishery extension work--India--Dakshin Dinajpur (District) 4. Fishery extension work--India--Uttara Cabbiśa Paragaṇā. 5. Fishery management--India--Dakshin Dinajpur (District) 6. Fishery management--India--Uttara Cabbiśa Paragaṇā. I. Mukhopadhyay, Swapan Behari, author. II. Dana, Shyam Sundar, author. III. Title.

SH107.G67 2017 DDC 639.3095414 23

Published by : **Scholars World**
A Division of
Astral International Pvt. Ltd.
– ISO 9001:2015 Certified Company –
4736/23, Ansari Road, Darya Ganj
New Delhi-110 002
Ph. 011-4354 9197, 2327 8134
E-mail: info@astralint.com
Website: www.astralint.com

Digitally Printed at : **Replika Press Pvt. Ltd.**

— *Dedicated to* —

All the fishery scientists and technologists of different disciplines whose publications form the very basis of this book

Preface

This is a comprehensive book on fisheries extension and management in which the theoretical aspects and research study has been explained lucidly. This shall serve as a handbook for extension workers in Govt. departments and in other organisation connected with fishery and rural development.

At the outset, the authors wish to cordially thank Prof. (Dr.) Debasis Majumdar, formers Vice Chancellor, Uttar Banga Krishi Viswavidyalaya and Dept. of Agril. Statistics, BCKV for his valuable guidance, and cooperation in entire statistical analysis of the present study.

We wish to express my sincere regards to Prof (Dr.) Atanu Sarkar, former Director of Extension Education, Uttar Banga Krishi viswavidyalaya for their unreserved help, active cooperation, encouragement and advice during the entire course of the present study.

We put on record my profound gratefulness to *Hon'ble Vice-Chancellor, Visva-Bharati University; Hon'ble Vice-Chancellor,* Uttar Banga Krishi Viswavidyalaya for their kind cooperation and timely help rendered by them during the course of investigation.

We acknowledge with thanks the flawless service rendered by Prof. D. Bhattacharya, Head, Department of Agril. Extension, Agril. Economics and Agril. Statistics (EES) and Prof. K. Boral, the Adhyaksha, Palli Siksha Bhavana. Deputy Registrar (Academic and Research), Visva Bharati and all the teachers and staffs members of Department of Agril. Extension, Agril. Economics and Agril. Statistics (EES) during the entire course of work. We ever grateful to all of them and owe our special thanks for their broad heart.

A deep sense of indebtness and appreciation are due to Shri Hironmoy Chakravarty for his erudite effort in modifying the manuscript.

We are especially indebted to Dr. Narayan Chandra Sahu, P.C, Sasya Shamala KVK, Dr. Rabin Sarkar, Ex-Professor Katwa College and Shri Baglapada

Bhattacharya, Ex Head master, Katwa Kashi Ram Das Institution for their all round help, valuable suggestions and inspirations. We are also thankful to Dr. Indranil Das for his encouragement and affection throughout the study period.

We are grateful to Dr. Sankar Saha, Mr. Sibananda Singha Mr. Bappa Paramanik, Mr. Siddikul Islam, Dr. Sukanta, Biswas, Mr. Nakul Mondal, Dr. Soumen Mahapatra, Mr. Gorky Sarkar, Mr. Sudipta, Debnath for their constant support, affection and timely help.

The help of Mr. Biswajit Talukder (Bappa) for the arduous task of Computer typing of the manuscript are acknowledged with complete satisfaction for various untiring services.

Words are not ample enough to express our sincere gratitude and thanks to Smt. Hena Mukhopadhyay and Smt.Mousumi Dana.

We find, in our conceit, We cannot formulate any words that would express our *ad- infinitum* indebtness to Shri Purnendu Sekhar Goswami, Smt. Kalpana Goswami, Smt. Archana Goswami, Smt. Kakali Goswami, Shri Giri Gobardhan Goswami, Smt. Tapasi Goswami, Shri. Ramanuj Goswami, Smt. Pompi Goswami, Shri Krishnarjun Goswami, Smt. Tuhina Goswami, Dr. Biswarup Goswami, Smt. Jyotirani Goswami, Miss. Swaranika Goswami, Shri. Soumyabrata Goswami, Shri. Debartha Goswami, Shri. Subhrojit Goswami, Shri. Nilanjan Goswami, Miss. Oindrila Goswami, Adrija, Anandi and Anika for their silent prayer for success.

The patience, silent service and excellent cooperation kindly extended by Smt. Sanghamitra Goswami during the several years of writing and updating of the manuscript of the book is gratefully acknowledged. We hope the book would be useful to students, researchers, planners, farmers and all stakeholders in fisheries extension. This is the first attempt in our endeavour to provide comprehensive information in the areas of fisheries extension.

No words are enough to express my heartiest gratitude to the respondents of this study for their help, kind cooperation during data collection.

Dr. Biswajit Goswami

Prof. (Dr.) Swapan Behari Mukhopadhyay

Prof. (Dr.) Shyam Sundar Dana

Contents

Abbreviations

AH: Animal Husbandry
AI: Artificial Insemination
AICPND: All India Coordinated Project for National Demonstration
AICRP: All India Coordinated Research Project
ANOVA: Analysis of Variance
APL: Above Poverty Level
BAES: Bureau of Applied Economics and Statistics
BCKV: Bidhan Chandra Krishi Viswavidyala
BDO: Block Development Officer
BFDA: Brackishwater Fish Farmers Development Agency
BPL: Below Poverty Level
CDP: Community Development Programme
CIBA: Central Institute of Brackishwater Aquaculture
CIFA: Central Institute of Freshwater Aquaculture
CIFE: Central Institute Freshwater Education
CIFRI: Central Inland Fisheries Research Institute
CIFT: Central Institute of Fisheries Technology
CMFRI: Central Marine Fisheries Institute
DRDA: District Rural Development Authority
DWCRA: Development of Women and Children for Rural Area
FAO: Food and Agricultural Organisation
FEO: Fishery Extension Officer
FFDA: Fish Farmers Development Agency

GDP: Gross Domestic Product
HYVP: High Yielding Variety programme
IAY: Indira Awas Yojana
ICAR: Indian Council of Agricultural Research
ICDP: Intensive Cattle Development Project
IRDP: Integrated Rural Development Programme
IVLP: Institution Village Linkage Programme
KVK: Krishi Vigyan Kendra
LLP: Lab to Land Programme
MFAL: Marginal Farmers and Agricultural Laborers Programme
NA: Non Adoption
NABARD: National Bank for Agriculture and Rural Development
NAIP: National Agricultural Innovation Project
NATP: National Agricultural Technology Project
NCA: National Commission on Agriculture
NES: National Extension Service
NGO: Non Governmental Organization
ORP: Operational Research Project
PA: Partial Adoption
RLEGP: Rural Landless Employment Gurantee Programme
RB: Rice Bran
RCD: Raw Cow Dung
SFDP: Small Farmers Development Programme
SFSDC: State Fish Seed Development Corporation
TOT: Transfer of Technology
TRYSEM: Training of Rural Youth for Self Employment
UBKV: Uttar Banga Krishi Viswavidyalaya
WHO: World Health Organisation

List of Figures

List of Tables

Chapter 1
Introduction

Kautilya's Arthasastra one of the oldest treatises dealing with a number of sociological subjects indicates that fish culture activity in India dates back to more than 2000 years. Perhaps it started when the human settlement moved away from the river banks to the hinterland. Paddy fields and low lying areas in flood plains and those connected to the estuaries and estuarine creeks become the cradle of aquaculture, where inundation, caused either by monsoon rain or tidal water, brought the natural seed of fin fish and shell fish, which got automatically trapped after the water receded. The observation of this phenomenon eventually gives rise to the technique of 'trapping and holding' fish seed and raising them to table size, and thus this marked the beginning of aquaculture in India. Over the ages geometric increase in population has resulted in relative shortage of food from the land resources and land : man ratio is decreasing day by day. In this context, it is of no little importance that the fish crop of animal protein is among the largest of foodstuffs being produced presently. Its relative importance as a source of animal protein too is considerable. Moreover, the cost of production is also less in fish culture compared to that in land-based animal. For these reasons people are being attracted more and more to these types of culture practices.

Fish has nearly ever been occupying an important place in the global food basket as a relatively cheap source of animal protein with higher consumer acceptability. Fish and fisheries have gained considerable importance both as a source of nutrition and as a source of steady and growing income at various levels - individual, state and national. Coming to the Indian scenario, the fisheries sector occupies a unique status in the national economy and provides employment opportunities to a significant section of the growing population in the country. In the last 50 years fish production has increased substantially with a corresponding reflection on a rising trend of its contributions to the gross domestic product (GDP) and the country's export earning.

Notwithstanding all endeavours at stepping up the supply of food stuff, over the ages, the relatively fast pace of the increase in population has tended to overtake the rate of increases in food production thereby resulting in virtual shortage of food from the land resources and a corresponding decline in the land : man ratio day by day. In this context it is of no mean importance that the fish crop of animal protein figures as one of the largest of food stuff being produced presently. Its relative importance as a source of animal protein too is considerable. Moreover, the cost of production is also less in fish culture as compared to that of rearing land based animals. No wonder an increasing human population coupled with a correspondingly rising demand of animal protein in the country is exerting enormous pressure on the fisheries resources available for current exploitation and potential utilization.

Many millions of people around the world find a source of income and livelihood in the fisheries and aquaculture sector. The most recent estimates indicate that 58.3 million people were engaged in the primary sector of capture fisheries and aquaculture in 2012. Of these, 37 percent were engaged full time, 23 percent part time, and the remainder were either occasional fishers or of unspecified status (FAO, 2014)

Most of this growth has taken place in the 1980s and that too mainly in the Asian countries, which accommodates four fifth of the fish farmers of the world dwelling. In India, the importance of aquaculture sector is well known as a provider of livelihood to seven million aquaculturists, the prime mover of the coastal economy and a major foreign exchange earner. Development of several suitable technologies coupled with extension activities during the last five decades has pushed Indian fish production manifold, particularly to become the second largest aquaculture fish producer in the world. The fish production has increased from 0.75 million tonnes in 1950 to 9.72 million tonnes at present, a more than twelve fold increase (Ministry of Agriculture, 2015). At the same time the share of inland fisheries has gone up from 29 percent to over 60 percent with production of over 6.14 million tonnes from the inland fisheries sector, the country has occupied the second position in the world (Ministry of Agriculture, 2015). Presently the contribution of aquaculture to the GDP of the country is estimated at 1.1 percent (NFDB, 2016). The unparallel average annual growth rate of over 4.5 per cent over the years has placed the country on the forefront of global fish production, only after China. The present fresh water aquaculture production of about 2.9 tonnes/ha, which constitutes about 95 per cent of the total aquaculture production and about 40 per cent of the total fish production of the country, has been made possible by the phenomenal growth in development of carp farming in small inland water bodies, that is ponds and tanks (Handbook of Fisheries and Aquaculture, 2013, ICAR publication, India). Even then there still remains a vast potential waiting to be trapped. Besides meeting the domestic needs, the dependence of over 14.5 million people on fisheries activities for their livelihood and foreign exchange earnings to the tune of US$ 3.51 billion (2012–13) from fish and fisheries products, amply justifies the importance of the sector on the country's economy and in livelihood security.

The country is endowed with vast and varied resources possessing river ecological heritage and rich biodiversity. India possesses vast inland aquatic resource in the form of 1,91,024 km of rivers and canals, 3.15 million hectares of reservoirs,

2.35 million hectares of ponds and tanks, 1.3 million hectares of oxbow lakes and derelict waters and 1.24 million hectares of brackish water and estuaries 0.29 million hectares (NFDB,2016).

It has been estimated that about 0.8 million tonnes of inland fish is contributed by different types of inland open water systems. Though the production details of these water bodies are not available, it is believed that capture aquaculture production from rivers and estuaries contributes only a small share of the total inland catch and that the bulk of the production comes from reservoirs and flood prone wet lands which are managed on the basis of culture based aquaculture or various other forms of enhancement (Ayyappan and Jena, 2003).

With a production of over 3.4 million tonnes from the inland fisheries sector, India has come to occupy the second position in the world, next only to China (NFDB, 2016). Such growth has been possible due to impressive development in aquaculture, mainly through carp based fish farming in fresh water tanks/ponds and shrimp farming in the coastal areas. In addition to contributing to more than 35 per cent of the per capita fish consumption, it is generating employment in the rural sector and incidentally also arresting irrigation. Besides the traditional waterway resources of ponds and tanks, appropriate technologies have made it possible to utilize reservoirs and canals too for aquaculture. Marine fisheries have progressively increased by nearly six times during the past 55 years recording 3.0 million tonnes in 2014-15. During the nineties the compound growth rate of marine and inland fisheries was respectively to the tune of 2.2 and 6.6 per cent per annum.

It can be observed that fish production from marine capture fisheries is getting stagnated. As a result a high degree of pressure has been exerted on the ongoing process of the intensified development of fresh water and brackish water resources through scientific aquaculture practices popularization of suitable technologies coupled with adequate extension activities and the introduction of culture oriented fisheries activities in existing large number of small tanks and ponds at the behest of the Central and State Govt. All these developments have led to a tremendous boost to inland fish production earning for India recognition as one of the major producers of inland fish through aquaculture.

Fish culture of single species is giving way for composite fish culture which involves stocking a selected mix of various fish species, providing adequate and timely feed and manures, ensuring watch and ward and organizing, harvesting and marketing. In pursuance of the recommendations of the Technical Committee on Inland Fisheries to popularize the technology of composite fish culture as evolved with adequate scientific research and trials by the Central Inland Fisheries Research Institute, Barrackpore, a new institutional framework in the form of Fish Farmers Development Agencies (FFDAs) started functioning from early 1973-74 in the selected states. The FFDAs were established with the main objective of serving as "nucleus centres" to provide the much needed mechanism for promoting scientific fish farming as well as to coordinate the role of various institutions for this purpose. The specific objectives of the FFDAs include (i) to introduce and popularize improved fish farming practices (ii) to bring all suitable fresh water resources like ponds, tanks and other water sheets under fish farming (iii) to build up a cadre of trained fish farmers who will take up fish farming, and (iv) to contribute to the strengthening of the rural economy by providing additional occupation, thereby also enhancing additional income.

The phased growth in the establishment of FFDAs in different states has resulted in the establishment of a network of 422 FFDAs functioning in all potential districts of the country. During 2000-01 about 30,000 ha of water area was brought under the fish culture and about 20,000 fish farmers were trained in improved fish farming practice under FFDA programme in the country. It is estimated that only about 20 per cent (0.46 million ha) of the available resources have been utilized so far by the Fish Farmers' Development Agencies (FFDAs). The average yield from ponds which was of the order of 600 kg/ha/yr in 1974 has gone upto 2200 kg/ha/yr by 2001-02 (Tripathi, 2003). What is required to be set up as the target of development effort is an increase in the average yield as well as production per unit area. It could be achieved by emphasizing two objectives :

1) To improve the socio-economic condition of the aquaculturists, especially the backward population amongst them;
2) To develop a positive attitude amongst the rural aquaculturist towards scientific aquaculture in the interest of inducing them to adopt the innovations or new technology in the field of aquaculture sector.

In the Tenth and Eleventh Plan periods, fish production registered a steady growth and it is expected that this growth would continue in the Twelfth Plan also, albeit with appropriate incentives and checks and balances. With the scenarios for inland and marine fisheries described in the foregoing paragraphs, the fisheries sector is likely to grow around 6.0 percent on an average per year during the Twelfth Five-Year Plan. With this growth rate, the total fish production is targeted at 11.58 mmt by the end of the Twelfth Plan Period (2016-17). To achieve this growth, it is estimated that the sector would require Rs. 6 000 crores for the Twelfth Plan period. The proposed programmes for HRD, institutional strengthening, policy reforms and overall improvement in the management and governance aspects during the Twelfth Plan period will also have spill over positive impacts on the output and performance of the fisheries sector during the subsequent Plans (Planning commission, 2012).

The main thrust in freshwater aquaculture in the country has been attributed to the organized activities of the Fish Farmers' Development Agencies (FFDAs), set up in mid 70's. Of the total 422 FFDAs sanctioned to the states, about 400 FFDAs are operational mostly at the district level and some at the regional/state level. These agencies provide technical, financial and extension support to the fishermen/fish farmers. These agencies have played a pioneering role in the promotion of fresh water aquaculture and have been instrumental in bringing about 5.30 lakh ha water area under scientific fish culture inducing farmers to take to improve practices. The most significant achievement of the agencies has been the building up of a cadre of well trained fish farmers and entrepreneurs in the country.

Coming to the conditions obtaining in this regard in the state of West Bengal, we are to note at the outset that considering historically, geographically and strategically, West Bengal has been playing a rather significant role in regard to fish culture since long past. West Bengal is one of the important maritime states in the country, located within 21⁰38′ to 27⁰10′N (Latitude) and 85⁰38′-89⁰50′E (Longitude) with an area of 87.853 sq. km. The state is endowed with both marine and fresh water resources. West Bengal possesses vast inland aquatic resources in the form of 2.76 lakh hectare of ponds and tanks, 0.41 lakh hectare of beel and boars, 0.16 lakh hectare of reservoir, 1.72 lakh hectare of river, 0.80 lakh hectare of canal and 2.10 lakh hectare of brackish water area. On the marine side, 158 km of coastal line and nearly 17049 km^2 of shelf area abounds both pelagic and demersal fisheries resources. The total fish production of the state during the year 2012-13 was 14.90 lakh tonnes with 1.52 lakh tonnes from marine sector and 13.38 lakh tonnes from inland resources (Hand book of Fisheries Statistics, 2012-13). The State incidentally, is again the highest consumer of fish and fish products with a per capita fish consumption being 15.6 kg as against the world average 18 kg and the national average of 8 kg. (Ministry of Agriculture, Govt. of India, 2014). The domestic supply being inadequate, a part of the demand has necessarily to be met up by procuring fish from other states. Even then, there remains a considerable gap in supply to cater to the demand of the fish eating population, which grows steadily each year with increase in population. The contribution of West Bengal to the total fish production of the country is about 18.28 per cent while the contribution of West Bengal to total consumption of fish food is about 28.57 per cent. West Bengal is on the top of the list in fish seed production in the country. The fish seed production of the state during 2012-13 was 15002 million numbers (Hand book of Fisheries Statistics, 2012-13).

A review of the statistical data pertaining to the total quantity of fish seed stock in the cultivable water of the State trends to reveal that there has in the recent past been a substantial increase in the quantity of fish seed stocked in the ponds as compared with earlier period (say, the nineties). This fact implies that fish culture in cultivable water has of late been gaining the importance of a commercial venture in the state. However, there still remains a tremendous scope to increase the fish production of the state by proper adoption of recommended scientific culture practices as also by bringing all the unutilized resources under the fold of scientific fish culture. This incidentally, suggests and inculcation of innovative scientific technologies and modern aquaculture practices.

In the year 2014, India ranks second in total fish production in the world (158 million tonnes) by producing 9.58 million tonnes of fish which is approximately about 6.06 per cent of the world's fish production. In inland fish production, India ranks third next to China by producing 6.14 million tonnes of fish which constitutes roughly 9.9 per cent of the world's inland fish production (Ministry of Agriculture, Govt. of India).

In course of the last three decades or so the Fish Farmers' Development Agencies (FFDAs) has been serving as a sort of nodal agencies for development of fish farming in its myriad aspects. A major national programme of Fish Farmers' Development Agencies (FFDAs) at the district level was initiated at 1973 to provide necessary administrative and infrastructural support, arrange training to beneficiaries, ensure mobilization of inputs and input support to fish farmers and also arranging institutional finance though bank credit. The FFDA was entrusted to bring all the fallow culture fishery resources under optimum fish production progressively. The agency played a leading role in extension of fish culture technology and also helped creating well-organized skilled and enterprising fish farmers communities (Sinha, 1999). As indicated in reports under Inland Fisheries Project, the FFDA has made substantial progress in all the districts of West Bengal. Pond fishery occupies here an important position amongst all the inland fishery resources. The production of fish from culture based fishery is not optimum in relation to the production potential of the state. So, it is high time that a study was undertaken into certain basic questions such as :

- ☆ What is the present knowledge level of the fish farmers about scientific fish culture technology?
- ☆ What is their attitude towards the scientific fish culture technology and what is the status of the level of adoption of the scientific fish culture technology?
- ☆ What are the different relevant characteristics of the fish farmers like personal, socio-economic, psychological, communication and extension character that affect attitudes and adoption behaviour of the fish farmers? And finally
- ☆ What are the prevailing constraints that inhibit their adoption of practices?

All these questions are expected to help to promote a better understanding of the present situation by bring in suitable changes in strategies adopted for better performance by focusing on the existing lacunae in aquaculture extension, which plays a crucial role in the matter of effecting changes in the fish farmers' scientific and technological knowledge of aquaculture with the ultimate aim of upgrading their existing operational practices and improving their farm management skill. Reorganization and improvement of aquaculture extension further seeks to fortify the overall process of the unutilized and under-utilized water areas under modern fish culture operation to the maximum possible extent. The competence and efficacy of the fish farmers through the inclusion of knowledge and skill leads to higher earnings by the fish farmers through improving the productivity and the profitability of their farming and a consequent improvement in their standard of living.

Statement of the Problem

Research and demonstration have shown that significant increases in the production of fish from the available fresh water culture resources could surely be achieved through the introduction of scientific fish practices on the wider scale. It is through the adoption of scientific fish culture practices in the existing field conditions that it has generally been possible in the recent past in bringing about phenomenal changes in productivity, for instance, from the earlier production rate of 600-1000 kg/ha/yr to the current fish yields ranging from 3,000-6,000 kg/ha/yr.

Research in aquaculture is a valuable input to increase and stabilize production. With this guiding principle in view the last three decades have made sustained research efforts, which resulted in the development of modern high-yielding production technologies. The results of the research which are found to be relevant for general adoption are tested for their feasibility and economic viability on the farmers' fields in the target area, before they are accepted for inclusion into modern technology, which is a combination of several tested practices comprising a 'package'. The package is then transferred to the farmers. An important precondition to a sound economy is a balanced growth of research and extension system. In fact, the two systems are complementary to each other and must go hand in hand. However, it is a pity that this important fact was not fully realized earlier. There is a wide gap between what could be achieved and what is being achieved on the farm. A big gap exists between available technologies and their rapid transfer to the farmer. It is estimated that only 20-25 per cent of the modern technologies developed are used under actual field condition in India (Sharma, 1988). Despite all the rich profitability of the technology of scientific fish culture practices the extent of, its spread has still remained insignificant due to various reasons. The present study has been designed to investigate some factors which closely related with the attitude and the adoption behaviour of fish farmers towards scientific fish culture, so that the knowledge could be used in rapid diffusion of the scientific technology to the fish farmers.

Keeping the above background and prospects in view, the present study was proposed to be undertaken entitled as "Factors affecting attitude and adoption behaviour of fish farmers towards scientific fish culture in two districts of West Bengal, India" with the following specific objectives:

Objectives of the Study

- ☆ To study the selected socio-economic, psychological and communication attributes of the fish farmers in the study area
- ☆ To ascertain the attitude and adoption behaviour of fish farmers towards scientific fish culture
- ☆ To determine the factors affecting the attitude and adoption behaviour of fish farmers with regard to scientific fish culture
- ☆ To bring out the relationships of the socio-economic, psychological and communication characteristics of the fish farmers with their attitude and adoption of scientific fish culture
- ☆ To find out the constraints in the adoption of scientific fish culture as experienced by the fish farmers

Scope of the Study

Attempts have been made in the present investigation to study the behavioural changes in terms of social, economic, technological and psychological correlates of the fish farmers. The study concerning the attitude and adoption behaviour of the fish farmers towards scientific fish culture is likely to provide valuable information in developing suitable strategies for implementing future fishery extension programme more effectively.

- ☆ The study is expected to generate a data base about the existing practices being followed by the fish farmers of West Bengal in the field of fish culture.
- ☆ The research would in all probability help fishery extension workers by furnishing effective guidance to overcome the negative attitude of fish farmers towards scientific fish culture.
- ☆ The present study involves an identification of the constraints experienced by the fish farmers towards scientific fish culture. As such these findings are likely to help the policy makers, administrators, extension personnel in effective implementations of the fishery extension programme and intervention by ameliorating the identified bottlenecks.
- ☆ Above all, the present investigation is expected to encourage further studies in other regions. Besides, researchers would find useful hints and suggestions in the nature of guidelines for future studies on this aspect.

Limitations of the Study

Every empirical study tends to be subject to certain limitations depending on the nature, scope and area of the study. The present study is no exception. An appreciation of the limitations of the study may also be said to be helpful in assessing the utility and justification of the study. The limitations are briefly stated below.

- ☆ Based on previous research work done and discussion with experts, 24 independent variables which are relevant, being directly or indirectly related to the attitude and adoption aspects were selected for this study as it is not possible to include all the variables.
- ☆ The findings of the study are based on the information given out by the respondents and the opinions expressed by them. Therefore, the possibilities of biases and prejudices can not be ruled out. The quality of the information is also subject to the ability of the respondents to recall.
- ☆ Since it is not physically possible for an individual researcher to extend his study to a larger area, the study was confined to two districts of West Bengal. Hence, the findings emanating from the study may be applicable to this area or the other areas marked by identical conditions.

Layout of the Study

A total of seven chapters constitute this study. The first one being the 'Introduction' providing background information, statement of the problem, scope, objectives of the investigation along with its limitations. The second chapter deals with 'Review of Literature', seeks to provide an insight into similar types of studies carried out by the researchers in the past and an attempt has been made to convey an understanding of technical terms and concepts, etc, related to the study in their theoretical perspective. The third chapter concerns with 'Research Setting' deals with various information regarding socio demographic background and resource availability of the people and area under study. The fourth chapter *i.e.* 'Research

Methodology' provides a description of the sampling design, various variables and their measurement, techniques of data collection, and statistical analysis. The findings of the study along with discussion there upon are presented in chapter five, entitled as 'Findings and Discussion'. The sixth chapter is devoted to 'Summary and Conclusion' outlines the summary of and the conclusion based on the result of the study and also spelling out the implications. 'Bibliography' and 'Appendices' are provided at the end of the study.

Chapter 2
Review of Literature

A comprehensive review of relevant literature is essential in that it provides a sound base for carrying out any research study. The major purpose of the review of literature is to organize and elucidate one's acquaintance with the entire gamut of the past and contemporary research in the particular field of investigation. It aims to provide in a nutshell information on 'what has been done' as well as indicate in the light of that knowledge, guidance to 'what is to be done'. The other aims of the exercise are-

- ☆ To assist in the delineation of the problem area
- ☆ To provide a theoretical framework and guidelines for deriving hypotheses
- ☆ To propose an insight into the methods and procedures to be adopted
- ☆ To suggest operational definitions of the major concepts to be used, and
- ☆ To build up support for the logical interpretation of the findings of the present study

In view of the dearth of illuminative literature on Fisheries Extension, it is proposed to step out a bit and seek to build up an alternative frame work of a more or less comprehensive nature by drawing on literature relating to studies on similar other fields like agriculture, animal sciences, poultry and allied enterprises.

The literature directly related to attitude and adoption behaviour of fish farmers with regard to scientific fish culture is rather limited. Accordingly, an attempt has been made liberally to draw on literature concerning studies on allied activities in farm sectors. The selected reviews in line with the objectives (and variables) of the study have in this chapter been presented under subheads as detailed below-

2.1 Conceptual orientation

2.2 Profile of fish farmers

2.3 Attitude towards agricultural and allied farm enterprises

2.4 Association of attitude with independent variables

2.5 Adoption of recommended practices of agricultural and allied farm activities

2.6 Association between adoption of recommended practices with independent variables

2.7 Constraints to adoption of scientific fish culture

2.1 Conceptual Orientation

The development of a conceptual framework makes a research more meaningful, for, whatever, an empirical investigation without theoretical basis inevitably becomes bogged down in masses of irrelevant data and at the same time ignores potentially fruitful objectives (Rogers and Svenning, 1969). The chapter accordingly, deals with the conceptual orientation of the study in question under the following subheads:

A. Concept of scientific fish culture- history, development
B. Dependent variables
C. Independent variables, and
D. Constraints to adoption of scientific fish culture
E. Conceptual model

A. Concept of Scientific Fish Culture: History, Development

In India fish and fisheries have gained considerable importance both as a source of nutrition and as a source of steady and growing income at the individual. Traditional fish farming is an age-old practice in India. In West Bengal, Catla, Rohu and Mrigal were reported to be stocked in ponds in the ratio 3: 4: 4 (Alikunhi, 1957). This has progressed from stocking from single species to multi species culture commonly known as *composite fish culture* which was significant turning point in the development of inland fisheries in the country. An important land mark was achieved through the introduction of Chinese silver carp and Grass carp during 1959.

A system of polyculture comprising of Indian major carps, common carp and selected Chinese carps introduced in India has come to be known as *composite fish culture*. The main objective of composite fish culture system is to obtain high production per hectare of water body by stocking fast growing compatible species of fish of different feeding habits. Under this system of culture, most of the available ecological niches of the pond ecosystem are effectively exploited by culturing compatible fish species of different feeding habits. It is basically a six species culture *viz.* catla, rohu, mrigal, silver carp, grass carp and common carp which graze from different zones of the pond.

The basic principle of composite culture is that when compatible species of different feeding habits are stocked together, they secure for themselves in the most efficient manner all life requisite available in the pond for fish production without harming each other (Jhingran, 1997). It is possible to achieve fish yields ranging from

2000-10000 kg per hectare depending on package of practices followed in extensive, semi-extensive and intensive composite fish culture systems. The important management practices adopted in composite fish culture are described below:

1. Pre-stocking Practices

1. Before fingerlings are stocked in culture ponds, it is essential to clear them of aquatic vegetation, predatory and weed fishes and excessive silt (if present at the pond bottom). Manual method may be employed if the vegetation is scanty and scattered. Effective weedicides may be used to clear thick/dense aquatic vegetation.
2. Removal of predatory, weed and undesirable fish fauna before stocking of fingerlings is very much essential. Repeated netting or de-watering of the ponds helps in partial eradication. Use of pesticides is recommended for total eradication. Care should be taken to avoid subsequent entry by screening the inlets.
3. In order to enhance the natural fertility of pond/tank, it is necessary to manure the ponds with organic and inorganic fertilizers in recommended dosages. Timely pre-stocking manuring ensures availability of adequate natural fish food organisms. Further, liming before application of manures is also recommended which ensures desirable hydro-chemical parameters in the water body.

2. Stocking Practices

☆ Ponds/tanks may be stocked with fingerlings of recommended varieties of different fish species. The recommended number is 4000-5000/ha for village ponds and tanks, and it may go up to 10000/ha or even more in intensive culture systems.

3. Post Stocking Practices

☆ To get higher fish production, it is necessary to adopt supplementary feeding to ensure adequate availability of food. In addition, timely application of manures and fertilizers also to be practiced which helps in maintaining availability of natural fish food organisms during growth period. Recommended dosage and methods of supplementary feeding results in higher fish production besides reducing the wastage of feeds.

☆ Water quality and fish health management during culture period needs due consideration. Monthly netting operation reveals the growth and health condition of fish seed stocked. In about 10-12 months, individual fish are expected to attain 700-1000 gms or even more. Timely harvesting yields maximum returns. Partial harvesting of grown up fish and replacing with young ones may help in enhancing production.

History

Undoubtedly the fish culture system must have been developed thousands of years ago in India. Khona, the daughter-in-law of Varahmihir recommended the cultivation of vegetables in the embankment of fish ponds. In Kautilaya's Arthasasthra one of the oldest Indian treatises written somewhere between 321 to 300 BC indicate regarding fishes in a reservoir poisonous in times of war (Hora and Pillay, 1962).

This indicates that fish culture flourished at the time even in reservoirs. Another, document which describes methods of fattening fish in ponds is the encyclopedia of King Somesvara, Manasoltara, compiled in 1127 AD (Hora, 1951). Though an age old industry, mentioned as in 1127 AD, fish culture has been mainly confined to the eastern states of Bengal, Bihar and Orissa. Here great potentialities for a fish trade have always existed, fish being the main diet of the rice eating population of this region (Day, 1873, De, 1910). As the culturable carps (mainly catla, scientific name *Catla catla*) rohu (*Labeo rohita*), mrigal (*Cirrhinus mrigala*) and calbasu (*Labeo calbasu*) commonly called the major carps did not breed in confined waters, the fishermen devised effective methods of collecting their spawn from the flooded rivers, the Ganga, the Mahananda, and their tributaries during the monsoon months. Others explored ways of breeding them in confined areas by creating the flood, conditions artificially and were successful in their attempts. Such ponds, where carps were made to breed, were designated as bundhs. Large scale mortality as a common feature in egg stage, is found when reared in earthen pits in the vicinity of bundhs. The methods of transportation of spawn from the collection or production centre to the nearly ponds, or to markets also caused heavy mortality.

No attention was ever paid to growing the spawn in nurseries before stocking them in the larger water bodies, which resulted in poor yield. To many, fish culture consisted of purchasing some spawn from the market, putting them in the pond and finally reaping a harvest at the end of the year. In brief, though an ancient industry, fish cultural practices in India lacked a scientific background. The empirical methods as developed by local fishermen were often wasteful and faulty, thus seriously limiting production and returns. In the later half of the nineteenth century and early in twenty century attempts were however, made to culture exotic carps to develop the supply of fish in the areas where rich indigenous fish fauna is not found. Though gourami (*Ospheronomus goramy*) was introduced first in 1941 in Kolkata, it did not survive and was reintroduced in 1865 in China (Raj, 1916). Early in the twenty century, with the organization of fisheries departments in certain states, attempt were made to extend fish culture practices to other parts of India and to develop them on scientific lines. Notable amongst these are the efforts of Madras Fisheries Department according to the recommendation of Nicholson (1915) a large fish farm was established in the state Sunkesula (Krishna district) (Now in Andhra Pradesh) in 1911.

In subsequent years a large number of ponds, tanks, moats and irrigation reservoir were stocked with major carps fry and fingerlings. Further, an attempt to introducing exotic fishes, *O. goramy* from Java and Mauritius in 1916 (Nicholson, 1918, Hornell, 1920) at Sunkesula and *Cyprinus carpio* from Ceylon to Ootacamund in 1939, met with some success. Experiments were started since 1918-19 to determine the relative values of different kind of artificial food, such as groundnut and gingelly oil cake (Hornell, 1920) and a district by district survey of irrigation tanks and other inland waters were made. From 1931 onwards, the Madras fisheries department made initial attempts to acclimatize and grow the white mullet or milk fish, *Chanos chanos*, in various departmental ponds, the fry being plentiful on both the east and west coast during May-June and to a limited extent in November (Ranganathan, 1961). Unfortunately proper management techniques with regard to the culture of mullets and *chanos* have not far been developed. Other states followed Madras in matters of fishery research as fisheries departments were established, but little headway in fishery research was made. However, fairly exhaustive surveys for the development of fisheries were conducted in Bengal (Gupta, 1908, Naidu, 1939), Punjab (Dunsford, 1911), Uttar Pradesh (Eyde, 1923), Baroda (Moses, 1947), Mysore (Bhimachar, 1942) and Hyderabad (Rahi Mullarh, 1944). In Bengal improvements in carp cultural, methods were suggested by Sen (1941) and Hora (1943a-f, j,i, 1945d) in a series of articles to popularize and bring pond culture on scientific lines.

Development of Scientific Fish Culture

Composite fish culture which was developed at the Central Inland Fisheries Research substation at Cuttack in the mid sixties, has come a long way since its inception through the efforts of various research and extension projects of the ICAR, Central and State Govt. and the Agricultural universities. All India Co-ordinated Research Project on Composite Fish Culture of Indian and exotic fishes (The project is re-designated as All India Coordinated Research Project on Composite Fish Culture and Fish Seed Production – AICRP on CFC and FSP) was initiated in 1971 by the ICAR, with its headquarters at the CIFRI with a view to test semi-intensive fish culture in different parts of the country. Under the project, six centres one each in Andhra Pradesh, Haryana, Madhya Pradesh, Tamil Nadu, Uttar Pradesh and West Bengal were initially established. The results of initial experiments conducted on composite fish culture at two centres *viz.* Gujarat fish farm (Uttar Pradesh), Kulia fish farm, Kalyani (West Bengal) revealed gross yields in the range of 2693-3210 kg/ha whereas fish yield at Kulia fish farm was 3232 kg/ha in 6 months (Sinha *et al.*, 1973). Further trials conducted at all six centres under the project amply demonstrated that it is possible to obtain production ranging from 2436-6521 kg/ha/year through fertilization and supplementary feeding under various agro-climatic conditions in six states. In one of the experiments at Kulia farm (Kalyani) grass carp exhibited record growth of 5.04 kg in one year (Sinha and Gupta, 1975). At the Karnal centre (Haryana) production as high as 6191-7332 kg/ha/8 months were obtained with six species composition (Sukumaran, 1976). At Pune centre, production over 3 tons/ha/year (Tripathi, 1979). Chaudhuri *et al.* (1978) reported an average production of 7.6 tons/ha/year by adopting high density composite fish culture (13320 fingerlings) where facilities for replenishing the water already existed. All India Co-ordinated Research Project on Composite Fish Culture and Fish Seed Production also initiated similar works at some centrally sponsored centres (which were under the technical control of the project co-ordinator and administrative control of the state) where productions were in the order of 4107 kg/ha/6 months (Kaushalyaganga centre), 3968-4248 kg/ha/14 months (Godhra centre, Gujarat), and 3578 kg/ha/9 months (Coimbatore centre), respectively. During the period from 1981-84 all together 44 experiments were conducted under AICRP on Composite Fish Culture of Indian and exotic carps at different centres in which fish production ranged from 2.3-<11 tons per ha/annum. Highest production achieved was 10482 kg/ha/11 months (Anon, 1984).

Based on the initial high productions by the AICRP on Composite Fish Culture and Fish Seed Production, Department of Fisheries, Govt. of West Bengal initiated a large-scale adoption of technology in the villages of thirteen districts during 1975-76 with 91 ponds to begin with, which was extended to 584 ponds. Similarly CIFRI also initiated transfer of Composite Fish Culture technology through a project on national demonstrations initiated in 1973 in West Bengal and Orissa. Demonstration with four species combination incorporating Indian major carps and common carps at the rate of 7500 fingerlings/ha resulted in production of 3829 kg/ha/year (Kumar *et al.*, 1979). Further, progress was made at pond culture division of CIFRI at Cuttack and production as high as 9389 kg/ha/year were obtained (Chaudhuri *et al.*, 1975). Thus, the application of modern methods of fish farming technology as developed by CIFRI set the stage for aquaplosion in the country.

CIFRI published the first 138 case studies on composite fish culture in 1976 which included 66 case studies relating to privately owned water bodies and 72 in governmental/institutional water bodies (Anon, 1976). Subsequent publication included 126 case studies on composite fish culture (Ranadhir *et al.*, 1979).

With a view to providing intensive extension support to development of composite fish culture, the Govt. of India initiated establishment of Fish Farmers' Development Agencies (FFDA) under the chairmanship of District Collectors from 1974-75. The first FFDA started functioning in 1973-74 in Mysore (Karnataka) followed by East Champaran (Bihar) in 1974. During 1975, two more FFDAs one in Burdwan (West Bengal) and Motihari (Bihar) started functioning (Srivastava *et al.*, 1993).

B. Dependent Variables

The scientific fish culture technology has been a major breakthrough in aquaculture production. The average production of fish through traditional culture system remains as low as 600 kg/ha/year in seventies but at present the production of fish in the fish farmers' ponds could be achieved at least 10 to 12 times through adoption of carp polyculture and scientific fish culture technology. The fish yields ranging from 3000-6000 kg/ha/year have generally been achieved through adoption of package of practices of scientific fish culture in the field conditions.

Under suitable management conditions the recommended fish species, which is the basic component of scientific fish culture technology, give high yield.

It is presumed that other thing remaining constant, the yield of fish in scientific fish culture will depend on level of adoption of recommended technology by fish farmer and attitude of fish farmers about this technology. The present study has been designed to relate the adoption behaviour and attitude of fish farmers in respect of scientific fish culture technology with some selected independent variables. In this context it may be worth-while to understand a few basic concepts of attitude and adoption behaviour of individual.

a) Attitude towards Scientific Fish Culture

Attitude is usually considered as consisting of three basic components – thinking, feeling and reacting. The first aspect pertains to belief, it involves the thoughts of a person about the topic in question. The second aspect is related to value, whether the person feels attracted, repelled or neutral regarding the topic in question. The third element is a predisposition to behaviour.

Allport (1935) defined attitude as a mental and/or neutral state of readiness, organized through experiences, exerting a directive and dynamic influence upon the individual's response to all objects and situations with which it is related.

According to English and English (1958) attitude is an enduring, learned predisposition to behave in a consistent way towards a given class of objects; a persistent mental and/or neutral state of readiness to react to certain object or class of objects, not as they are conceived to be. It is by the consistency of response to a class of objects that an attitude is identified. The readiness state has a directive effort upon feeling and action related to the object. Sherif and Cantril (1948) concluded, while studying the relationship of attitude to action, 'a predisposition to action' as an essential feature of an acceptable definition of attitude. They have stated the following criteria for attitudes.

i. Attitudes always imply subject-object relationship,
ii. Attitudes are formed in relation to objects, persons and values ...attitudes are not innate but, are formed as a result of individual's contact with the environment,
iii. Attitudes have affective properties of varying degrees,
iv. Attitudes are more or less enduring states of readiness.

According to Kelly (1967) attitude refers to the opinions and beliefs that a person holds regarding objects, people, activities and social policies and practices.

Munn (1969) stated that attitude are learned predispositions towards aspect of our environment. They may be positively or negatively directed toward certain people, issues or institutions.

Chitambar (1973) stated that attitude can be loosely defined as tendency to act in some way toward some object, person, situation or idea. In general it is a feeling of like, dislike, attraction, repulsion or ideas. Attitudes are essentially acquired are learned. Learning or acquisition of attitude may be conscious or unconscious.

According to Lindgren and Byrne (1975) attitude is a relatively stable positive or negative evaluation of objects, persons, ideas or events.

Kagan and Havemann (1976) defined attitude as an organized and enduring set of beliefs and feelings towards some kind of object or situation and a predisposition to behave towards it is a particular way. It includes (i) A cognitive element (Belief), (ii) An emotional element (Feelings), (iii) A behavioural element.

According to Coleman (1977) attitude is a predisposition to feel, think and act towards some object, person, groups or event in a more or less favourlable or unfavourable way.

Thorndike and Hagen (1979) stated that attitude relate to tendencies to favour or reject particular groups of individuals, sets of ideas or social institutions.

Atkinson *et al.* (1983) defined attitudes are likes and dislikes, affinities for an aversion to situation, objects, persons, groups and any other identifiable aspect of the environment, including abstract ideas and social policies.

Supe (1983) stated that attitude is the degree of positive and negative feelings associated with some psychological object towards which people differ in varying degrees. It is developed in three phases, firstly towards the object, then the effect connected with the object and finally the action that can be undertaken with respect to that object. Attitudes are not innate but are formed as a result of the individual contact with object and its environment.

b) Adoption Behaviour of Fish Farmer

Behaviour of an individual refers to anything the individual does. According to Parsons and Shils (1951) behaviour is oriented to the attainment of ends in situations by means of normatively regulated expenditure of energy. This conceptualisation of behaviour implies that (a) behaviour is oriented towards attaining ends or goals and other anticipated state of affairs, (b) it takes place in situation, (c) it is normatively regulated and (d) it involves expenditure of effort or motivation.

According to Bohlen and Beal (1960) behaviour is an unit act consisting of (1) the receipt of the stimulus, (2) the interpretation of the stimulus and the circumstances under which it is received and (3) a response of action.

Chitambar (1981) states that concrete form of acquired new knowledge, skills and attitudes through experience that provides direction to interact in the society is termed as behaviour.

Farmer's behaviour to adopt a particular refers to both overt and covert behaviour within action frame of reference (Weber, 1947).

Ramsey *et al.* (1954) contended that the adoption behaviour would constitute of two components – behavioural (adoption) and cognitive (knowledge) in terms of the individual situation.

Rogers (1962) conceptualized adoption as a decision to continue full use of an innovation as the process by which an innovation spreads through a social system. Five stages of adoption process according to Rogers are awareness, interest, evaluation, trial and adoption.

Deway and Humber (1966) suggested that human behaviour has a trinomial nature which involves three components – (1) man's biological heritage – including both that which is transimitted genetically and the modification of this heritage through experience, (2) environment with social, cultural and geographical aspects, and (3) acquired variables including the subjective and covert ideas, knowledge and feelings on one hand and the overt objectively observable factors on the other.

Rogers and Shoemaker (1971) conceptualized adoption as one of the consequences of the innovation – decision process. According to them, the innovation-decision process is the mental process through which an individual passes from first knowledge of an innovation to a decision to adopt or reject and to confirmation of this decision.

Adoption is a decision to make full use of an innovation as the best course of action available (Radhakrishnan *et al.,* 1977).

Review of past literature in extension research furnishes several indices or procedures to measure adoption, which range from relatively crude to sophisticated indices, which take a variety of dimensions of adoption process.

Chattopadhyay (1963) developed adoption quotient scale which considered mainly the behavioural adoption. According to him the adoption quotient is a ratio scale designed to quantify the adoption behaviour of an individual. He considered potentiality, extent, time and consistency to arrive at various practices, according to their difficulty of adoption.

Bose (1965) assigned weightage to adoption of an improved practice in relation to the time of adoption.

Sengupta (1967) assigned each and every recommended practice with equal weightage with one and zero score on adoption and non-adoption of a practice respectively. He measured general adoption level by computing adoption quotient by dividing the number of practices and multiplied by hundred.

Supe and Singh (1973) brought into focus the concept of partial adoption. They have assigned additional weightage to divisible practices depending upon their difficulty. The adoption score was obtained by adding all the scores of all practices.

Subash Chandra (1986), while studying the consequences of adoption of improved fish culture practices by fish farmers has classified extent of adoption by fish farmers based on the percentage of adoption into low (0-25 per cent), medium (25-50 per cent), high (50-75 per cent) and very high (75-100 per cent). Similar procedure was earlier adopted by Meeran (1983) while studying the adoption behaviour of fish farmers in Tanjavour district of Tamil Nadu.

Balasubramaniam (1988) while studying the adoption behaviour of improved fish farmers in Tamil Nadu gave the weightages for individual practices and classified each practice into full adoption, partial adoption and non-adoption.

C. Independent Variables

Studies by Reiss (1952), Norris (1953), Beal (1958), Dean and Others (1958), Wilkening (1958), Thornton (1962), Graven (1963), Juliana (1964) and Singh and Singh (1970) have pointed out that human behavior is a multivariate phenomenon. In the present study it was intended to select a number of independent variables which could explain the attitude and adoption behaviour of fish farmers towards scientific fish culture.

Exploratory study, review of literature and discussion with progressive fish farmers, fishery scientists of Central Research Institute, Professors and experts of West Bengal University of Animal and Fishery Sciences, State Department of Fisheries and extension specialist helped in getting an idea of the variables relevant to the study.

From amongst, the large number of variables which may have theoretical relationship with attitude and the adoption behaviour of fish farmers in scientific fish culture, a manageable system of important variables that could be used to empirical measurement were selected. These were grouped as (a) socio-personal (b) socio-economic, (c) communicational (d) psychological (e) situational variables

(a) Socio-personal Variables

Certain attributes are the part of personality of human being as a socialized individual and they predispose him to act in a predictable manner towards social and psychological objects.

Age of Fish Farmer

Age refers to the number of years the respondent lived since birth at the time of interview and was rounded to the nearest whole number. Persons younger in age are likely to be progressive. Earlier studies indicated that age of the farmers was negatively associated with the adoption of practices (Ernest, 1973; Kapse, 1976 and Kumar, 1976).

Age of fish farmers may, therefore, have some relationship with attitude and adoption behaviour of fish farmers in scientific fish culture

Caste of Fish Farmer

A caste system is one whereby a society is divided up into a number of self –contained and completely segregated units (castes), the mutual relations between which are ritually determined in a graded scale (Hutton, 1963).Caste still retains its pivotal position in the social structure of Indian villages. Studies by Roy *et al.* (1968), Kumar (1976) have shown the caste of the respondent had positive relationship with the adoption of practices.

A relationship between caste of fish farmers and attitude and adoption behaviour of fish farmers in scientific fish culture is expected.

Family Size

It refers to the number of members present in the individual fish farmer's family. Generally up to 5 members is regarded as a small size family and large size family constitute more than 5 members. Large size family which has more work force may be more conducive to better management of fishery enterprise.

It is presumed that the family size of fish farmers may have some relationship with the attitude and adoption behaviour of fish farmers in scientific fish culture.

Family Type

In the present study, it refers to whether there is single or joint family system in the fish farmer's family. A family was considered as single when it consisted of husband, wife and unmarried children. A joint family consisted of other blood relations also.

Family type is likely to be related with the attitude and adoption behaviour of fish farmers in scientific fish culture.

Education of Fish Farmer

Education is generally believed to have the effect of widening the mental horizon of a person and thereby predispose him to be receptive to new ideas. Beal and Sibley (1967) pointed out that the individual's ability to read and write and the amount of formal education he possesses will affect the manner in which the individual gathers data and relates himself to the environment.

Jaiswal (1965), Shankariah (1965) and Bose (1965) found positive and significant correlation between education and adoption of farm practices. Studies indicate that earlier adopters have more years of education than later adopters (Rogers and Shoemaker, 1971).

Education of fish farmers may, therefore, have some relationship with attitude and adoption behaviour of fish farmers in scientific fish culture.

(b) Socio-economic Variables

Socio-economic status in this study refers to the position of fish farmers or his/ her family occupies in comparison to others with respect to occupation, annual income, land holding and fishing equipment possession. Fish farmers with higher economic status are likely to have higher economic motivation and may also be capable of investing more in fish culture.

A relationship between occupation, annual income, land holding and fish equipment possession and attitude and adoption behavior of fish farmers in scientific fish culture is expected.

Size of Land Holding

It is the amount of land possessed by an individual which may include ponds also. Studies in adoption and diffusion have indicated that size of holding in most cases is positively and significantly related with the adoption of improved farm practices (Singh and Sohal, 1967; Thorat, 1969; Kumar, 1976 and Sinha, 1978). Size of holding is also an indicator of economic status in rural India.

A relationship between the size of holding of fish farmers and attitude and adoption behaviour of fish farmer towards scientific fish culture is expected.

Social Participation

Social participation refers to the voluntary sharing in person to person and in group to group relationship beyond the immediate household (Hay, 1951). It shows the degree to which the fish farmers are involved in formal organisations as member or office bearer and attendance to meetings. Past studies (Jaiswal,1965 ;Reddy and Kivlin,1968 ; Earnest,1973 and Kapse,1976) have revealed a positive association between the extent of farmers social participation and level of adoption of innovations.

It refers to the degree to which a fish farmer was associated with different organization like, youth club, panchayat, co-operative etc. Social participation may, therefore, have some relationship with the attitude and adoption behaviour of fish farmers in scientific fish culture.

(c) Communication Variables

Communication is the process by which messages are transferred form a source to one or more receivers (Rogers and Svenning, 1969). Simon (1957) pointed out that action of an individual was a function of his exposure to the sources of information. He further defined exposure as the willingness of a farmer to react in a particular information media.

Past research studies (Beal,Rogers and Bohlen, 1956; Beal, 1958; Copp,1958; Lionberger, 1960;Leagans,1961; Wilkening, 1962 Singh and Jha, 1965; and Sinha,1978) indicate that the adoption of innovation is a function of contact with the information sources. Sources of information for the fish farmers may, in general, be classified into three categories—mass media, personal cosmopolite and personal localite. For a recent technology like scientific fish culture, mass media and personal cosmopolite sources of information appeared to be relevant and are considered for the present study.

Mass Media Participation

After reviewing a large number of studies Rogers and Shoemaker (1971) generalized that earlier adopters have greater exposure to mass media than later adopters.

A relationship between the use of mass media by the fish farmers and attitude and adoption behaviour of fish farmers towards scientific fish culture is expected.

Cosmopoliteness

Cosmopolite sources of information and conceptualized as those from outside the social system being investigated (Rogers and Shoemaker, 1971). Katz (1964) and Rogers and Shoemaker (1971) defined cosmopoliteness as the mobility of an individual out side his/her social system. They also observed that cosmopoliteness also reveals an openness to the rational evaluation of a proposed change and a willingness for contact with the world outside their communities. Personal cosmopolite sources of information play an important role in a programme of planned change. Personal cosmopolite sources are important because they act as carriers of change, provide face-to-face, word-of-mouth communication and motivate people to adopt the new technology.

A relationship between the utilization of personal cosmopolite sources of information by the fish farmers and attitude and adoption behaviour of fish farmers towards scientific fish culture is expected

(d) Psychological Variables

The framework of behaviour as discussed earlier, leads to assume that there are certain attributes that are a part of man as a socialised individual, that predispose him to act in a predictable manner. It may be assumed that adoption behaviour and attitude of fish farmers in scientific fish culture is not only the function of situational and socio-personal factors but is also influenced by the socio-psychological makeup of the individual.

The theoretical bases selection of the socio-psychological variables are discussed in brief.

Innovation Proneness

Rogers and Shoemaker (1971) defined innovativeness as 'the degree to which an individual is relatively earlier in adopting in adopting new ideas the other members of his system'. The norm of innovation proneness as viewed for himself is a matter of self- influence and the individual feels obligated to conform to his norm.

Innovation proneness of fish farmers has been defined as the behaviour pattern of individuals who have interest in and desire to seek changes in the techniques of fish farming and to introduce such changes into their operations when practical and feasible.

It is postulated that a fish farmer who views himself innovative prone is likely to change his attitude towards scientific fish culture and adopt the scientific fish culture practices.

Credit Orientation

It is difficult to break through the vicious circle of the tradition method of fish farming with low production and shortage of capital. For the average fish farmers this circle may be broken if new capital is introduced in the form of credit. Borrowing credit for commercial purposes pre-supposes and ability to view the future with confidence. The investment may or may not be justified in the light of actual production. But the willingness to take the risk is important and should come first.

A relationship may, therefore, be expected between credit orientation and fish farmers' attitude and adoption behaviour towards scientific fish culture practices.

Value Orientation

Values are related to the belief of an individual, but unlike beliefs, they involve an element of approval or disapproval.

Beal *et al.* (1957) defined values are the standards upon which evaluations are made, the criteria by which both ends and means are chosen. This definition implies than an individual is emotionally committed to the standards in such a manner that the standard influence, guide or direct his behaviour.

Kluckohn *et al.* (1965) stated that values are not goals but they are criteria by which goals are chosen and individual preference are ordered.

Nelson *et al.* (1965) observed values as systems specifying a code of conduct including means and goals.

According to Parsons and Shils (1965) value orientation refer to those aspects of the act or orientation which commit him to the observance of certain norms, standards, criteria, for selection whenever he is in a contingent situation, which allows him to make a choice.

A relationship may, therefore, be expected between value orientation and fish farmers' attitude and adoption behaviour towards scientific fish culture practices.

Risk Orientation

The term risk commonly refers to all outcomes which lead to losses deviations of realization from expectations (Heady *et al.*, 1937). Risk orientation may be described as the degree to which a fish farmer is oriented towards encountering risks and uncertainty in adopting scientific fish culture practices.

Studies in the more developed countries have shown that individuals vary widely in their willingness to take risk (Beal and Sibley, 1967. Jaiswal (1965), Ernest (1973), Ramachandran 1974) and Singh (1975) indicated that risk taking willingness is positively related with attitude and adoption behaviour of the farmers.

It is, therefore, expected that the fish farmers who are oriented to take moderate calculated risk will change the attitude of fish farmer to some extend and adopt the scientific fish culture practices.

Economic Motivation

Economic motivation may be conceptualized as the values or attitudes which attach importance to profit maximization as the ends and means. A fish farmer's economic motivation may be an indication of the degree of his willingness for investment of available potential resources in adopting innovations in fish farming. A fish farmer who views himself to be economically motivated may seek more monetary gains than a fish farmer with such traditional values as freedom from debt and self-sufficiency. Studies indicate that individual's economic motivation is positively and significantly correlated with adoption behaviour (Nair, 1969; Singh, 1969; Supe, 1969 and Singh, 1975).

A relationship between the economic motivation of fish farmers and attitude and adoption behaviour of fish farmers towards scientific fish culture is expected.

Knowledge

Knowledge is generally understood as an intimate acquaintance of an individual with facts. English and English (1958) defined knowledge as a body of understood information possessed by an individual and by culture.

Bloom (1958) defined knowledge as "Those behaviours and test situations which emphasized the remembering either by recognition or recall of ideas, materials or phenomenon".

Rogers and Shoemaker (1971), while dealing with knowledge of farmers with reference to adoption of innovation, divided knowledge into three types: 'awareness knowledge', 'how to knowledge' and 'principles of knowledge'.

Knowledge is one of the important components of behaviour and as such it plays an important role in the covert and overt behaviour of an individual. Past studies indicate that knowledge is positively and significantly related to the adoption of innovations (Moulik, 1965; Rao, 1968; Nair, 1969 and Ramachandran, 1974).

A relationship between the knowledge of fish farmer about scientific fish culture and attitude and adoption behaviour of fish farmers is expected.

(e) Situational Variables

The situation or the circumstances under which an individual receives a stimulus forms an important element in any theory of human action. This element was recognized by Parsons and Shils (1951) and they stated that behaviour of any living organism is called action only when it is analysed in terms of situation in which it occurs.

The importance of situational variables as predictors of human behaviour has been described by many social scientists like, Bohlen and Beal (1960), Kluckhohn *et al.* (1965), Newcomb *et al.* (1965) etc. These researchers pointed out that the behaviour of an individual is understood more in depth if one has the knowledge of certain situational variables which comprise the constructed world of reality within which an individual receives the stimuli and acts.

The theoretical bases for selection of the situational variables are discussed in brief.

Size of Water Body

Pond is an artificially confined body of water. Ponds are generally dug up for the purpose of irrigation or fish culture or both. The sources of water for ponds in most causes are rainfall and drained out water from the adjoining fields. In some rare cases, they may be connected with canals or larger water area.

According to Jhingran (1997) ponds are small and shallow bodies of quiet standing waters with only slight wind action, if any, in which extensive occupancy by higher aquatic plants is a common characteristic. No exact limits of area and depth have been laid down for a pond.

Size of ponds include banks and do not reflect the actual water area available for cultivation of fish. Further, the water area of a pond may vary from season to season but the size of pond as given in the land records remains constant.

As bigger ponds provide better opportunity for fish culture, particularly carps, some relationship between the pond size and attitude and adoption behaviour of fish farmers towards scientific fish culture is expected.

Source of Water

A dependable source of water supply must be available within or near the site even for undrainable ponds. However, unlike the drainable ponds, undrainable ponds require sufficient water to fill the ponds and to compensate the water loss through seepage and surface evaporation thereafter. The water supply to the pond as far as possible be natural, preferable rain water. However, alternative arrangements of the water supply should be made for dry season either from a deep tube well or irrigation canal or from perennial sources like spring, stream, river etc. ponds should be on the lower lands to allow accumulation of surface run of from a larger catchment area. Ponds fish production is influenced by the physical and chemical properties of the water. The chemical quality of water depends on its contents of dissolve salts. Water does not carry any dissolve salt. However it collects nutrient salt from the ground surface of the catchment area. The water should be neither too acidic not too alkaline, neutral or slightly alkaline water are most suitable for fish culture (Kumar, 1992).

Extent of Weed Infestation

Aquatic weeds are unwanted plants that grow within the water body and along the margins. Unlike in temperate climate, the pond fish culture in tropics faces serious problems due to weed infestation. They remove large quantity of nutrient from the water, which otherwise would go into the production of plankton growth. Even the poor fish crop is produced in weed chocked water which is difficult to harvest. The fishes are subjected to stress due to dissolve oxygen depletion and wide fluctuation between the dissolve oxygen values of the day and night. Decomposition of the dead aquatic weeds further creates the oxygen problem. Dense growths of submerged weeds restrict fish movement and interfere with fishing operation. Floating weeds very often cover the entire surface cutting off light drastically, thus resulting in critical reduction in primary production of the ponds (Kumar, 1992).

D. Constraints to Adoption of Scientific Fish Culture

According to Advanced Learner's Dictionary, the meaning of constraints is a thing that limits or restricts something, or freedom to do something.

The meaning of constraints is irresistible force; confinement; a restricted condition.

Waghmare and Pandit (1982) identified constraints in terms of causes hindering adoption of new technology.

Wagmare and Waghmare (1987) defined constraints as the item of difficulty faced by farmer in the process of technology.

Petharam (1985) and Zinyama (1988) called the problems and/or limitations as constraints.

Constraints in the present study indicates various inhibiting situations, encountered or expressed by the fish farmers.

E. Conceptual Models

The postulated relationship between variables is portrayed in conceptual model based on discussion and assumptions made earlier. The basic concepts included in the conceptual model is explained below:

The attitude of fish farmers towards scientific fish culture might be influence by number of factors (variable). An attempt has been made to analyse the extent of influence caused by these factors in influencing the attitude. Hence, the attitude of fish farmers was also considered as the dependent variable.

Earlier studies have significantly established the fact that the adoption behaviour of fish farmers towards scientific fish culture was influenced by socio-personal, socio-economic, psychological, communication and situational variables.

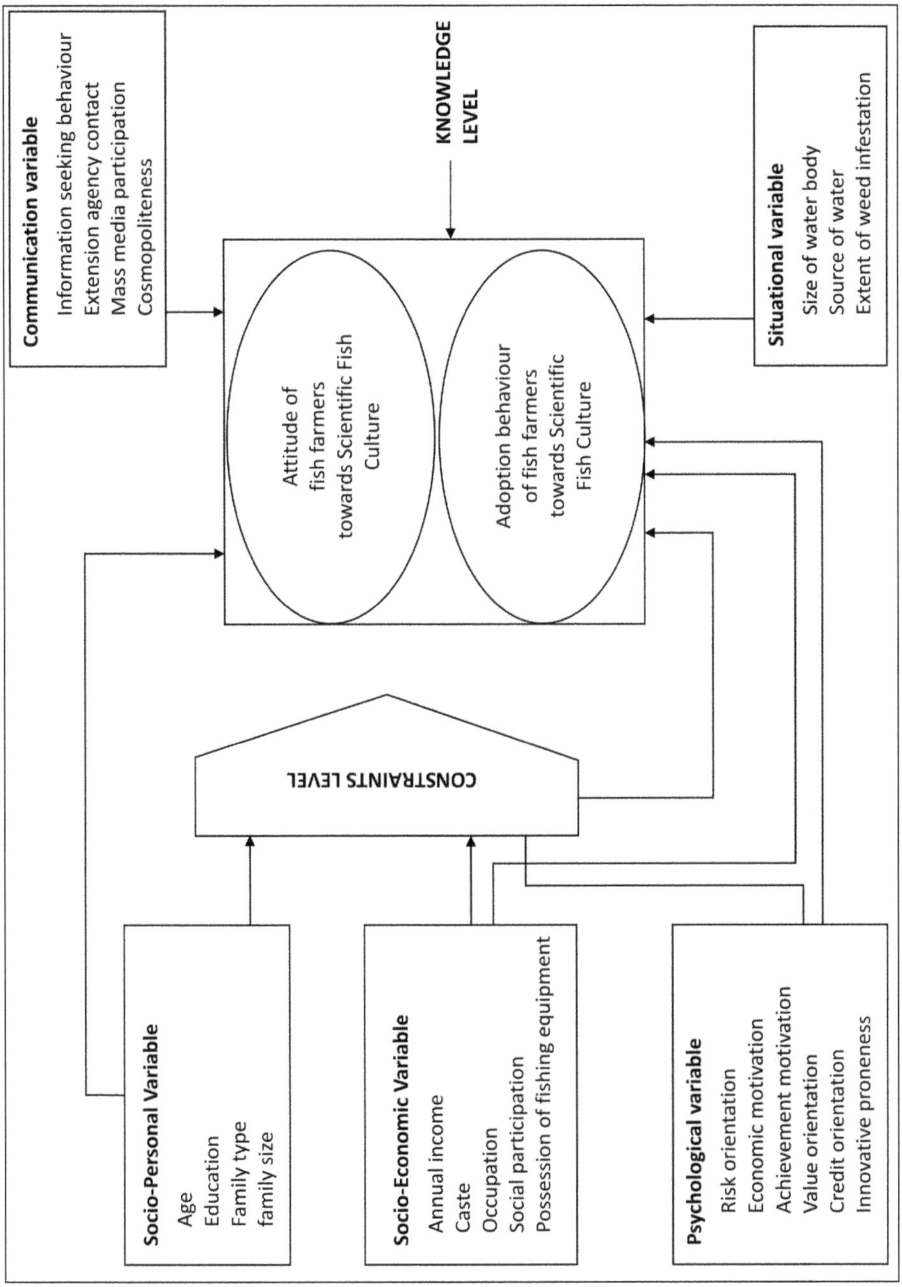

Figure 1: Conceptual Model of Attitude and Adoption Behaviour of Fish Farmers towards Scientific Fish Culture.

2.2 Profile of Fish Farmers

In the past, many researchers have studied the various socio-personal, economic, communication and psychological attributes of farmers in general and farmers involved in fish farming, dairying and other rural sector as high as 10148 kg/ha/year was obtained in the year 1979 (Mathew *et al.*, 1979).

Ponnappan (1982) in his study on fish farming programme in Tamil Nadu found that nearly 40 per cent of FFDA beneficiaries were educated upto primary and nearly ¼th of farmers were illiterate indicating low education status. Most of the farmers (63 per cent) belonged to young category. With respect to farm size, marginal and small farmers accounted for 51 per cent while 16 per cent were landless beneficiaries. The study also revealed that only 18 per cent of farmers belonged to high social participation while majority had medium and low social participation score. With respect to variables like social participation, innovative proneness and economic motivation more than 50 per cent of fish farmers belonged to medium category. Medium size fish farms were owned by majority of the fish farmers.

Meeran (1983) observed that majority of fish farmers belonged to young age group having primary to secondary level of education with non-fish culture as their major occupation. Fish farmers had better knowledge on fish culture practices. Most of the farmers belonged to low to medium level of social participation, extension agencies contact, economic motivation and scientific orientation.

Subhas Chandra (1986) in his study on consequences of adoption of fish culture practices in Kurnool district reported that majority of respondents had low education (primary to secondary), belonged to backward caste or scheduled caste with non-fish culture as their main occupation. Generally, the respondents possessed small sized fish farms. Most of the farmers had high to medium level of income and medium to low level of social participation, urban contact and material possession.

Misra (1987) reported that FFDA trained farmers are superior in their knowledge about improved fish farming practices than non FFDA fish farmers in West Bengal.

Venkataramanujam *et al.* (1987) in their study on socio-economic conditions of the chank fishermen community stated that all the adult females and 95.7 per cent of the males were illiterate. Regarding their housing condition, the study revealed 85 per cent had kuchha houses and 15 per cent possessed semi-pucca houses.

Balasubramaniam (1988) reported the socio-personal characteristics of fish farmers in Tamil Nadu. His study revealed majority (60 per cent) belonged to middle aged and nearly 50 per cent of them had education upto secondary followed by higher secondary (26.25 per cent). Nearly half of the fish farmers engaged in fish farming as a primary occupation, while 48.75 per cent engaged in fish farming as secondary occupation. Regarding annual income level, nearly 44 per cent of fish farmers had annual income below Rs.10,000 and 37.50 per cent had income level between Rs. 10,000 to Rs. 20,000.

Chede (1988) reported that more than 2/3rd of dairy farmers in Maharashtra (Amaravathi district) belonged to medium socio-economic status. About 33 per cent respondents did not possess land. Majority of respondents had medium level of scientific orientation, economic motivation and risk preference.

Rajukar (1988) observed that most of the fish farmers on Nagpur (Maharashtra) had low educational level, landless, poor with low economic and social status. Income from fish farming was also low.

Sujath Kumar (1988) revealed that most of the traditional fishermen belonged to middle (38.33 per cent) and old age (48.33 per cent) and among trawler owners most of them belonged to middle (43.33 per cent) and old (35.00 per cent) age groups.

Krishnaiah (1989) made detailed study of fish farmers' characteristics under FFDA programme in Andhra Pradesh. He reported that majority of fish farmers belonged to middle age group and almost 50 per cent of them had education upto primary school and 18 per cent were illiterate. With regard to occupation, 74.67 per cent of them possessed fish culture as main occupation and only 25.33 per cent respondents perceived fish farming as secondary occupation. Regarding initial status, he observed majority of the respondents had medium level of financial status while 27.33 per cent had low status and 9.34 per cent had high financial status. With regard to other variables like social participation, economic motivation, scientific orientation and risk orientation, majority of the respondents were of medium category.

Wayal (1990) in his study on members of Fisheries Cooperative Societies (FCS) found that fish farmers were mostly young, less educated, landless and were below poverty line with meager income from fish farming. Further they were deprived of information sources due to lack of extension work.

Govinda Gowda *et al.* (1991) conducted detailed investigation on socio-economic attributes of inland fishermen in Mandya district (Karnataka). The study revealed that nuclear family was the predominant (76 per cent) followed by joint family. Most of the fishermen (77.5 per cent) were illiterate and their membership in FCS was to extent of 74 per cent. Their communication exposure was very poor. Regarding occupation, 92 per cent of respondents had fishing as main occupation. Majority of respondents (65.30 per cent) annual income was below Rs. 5000.

Pounraj (1992) studied socio economic characteristics of inland fishermen in Tamil Nadu. He observed that most of the inland fishermen were middle and old aged with primary to secondary level of education. Most of the fishermen lived in tiled and terraced houses with medium to high level of income. Only 20 per cent of fishermen members lived in thatched houses. The other characters like innovativeness, extension agencies contact, material possession of inland fishermen were exhibited by the medium category.

Pramod and Sudhir (1992) reported wide variation in the education level of traditional prawn farmers in Orissa coast which varied from illiteracy to degree level.

Wayal and Ingle (1994) observed that most of the respondents in Akola district of Maharashtra were landless (69.39 per cent) followed by marginal (13.26 per cent) farmers. Gross annual income was below Rs. 6000 for 93.78 per cent of the respondents. Regarding educational level, 35.72 per cent of respondents were illiterate, 31.63 per cent primary and 22.45 per cent middle school and only 9.18 per cent were high school.

Beerannavar (1995) observed that 55.67 per cent of trained farmers belonged to medium knowledge category followed by 21.65 per cent and 22.68 per cent belonged to low and high level of knowledge with respect to improved dairy management practices in Dharwad district (Karnataka).

The study conducted by Mahandra Kumar (1996) revealed most of the FFDA fish farmers in Tamil Nadu were in medium age group and 70 per cent of the respondents pursued fish farming as secondary occupation and the rest as main occupation. Education level of farmers reveals 10 per cent illiterate, 15 per cent primary, 29 per cent middle, 15 per cent high school, 12 per cent PUC and 17 per cent graduates. Most of the fish farmers had medium income level (74 per cent) followed by 9 per cent in high group and 11.00 per cent in low income group. With respect to land holding 59 per cent of fish farmers belonged to medium category and 33 per cent to small category. His results also revealed that 59 per cent of farmers had medium knowledge while only 5 per cent belonged to low knowledge group and 38 per cent belonged to high knowledge category.

The study conducted by Pounraj and Sripal (1997) revealed majority (83.32 per cent) of inland fishermen of cooperative societies in Tamil Nadu (Thanjavur district) had primary to secondary level of education and were middle aged (75.83 per cent). He observed that about 68.52 per cent fall under medium to high level of income group and with respect to material possession majority of the fishermen belonged to medium category. About 65.83 per cent and 87.5 per cent of the respondents had medium level of innovativeness and extension agencies contact, respectively.

Munda *et al.* (1997) reported that majority of fisher folk in Orissa were middle aged (61 per cent) belonged to scheduled caste (60 per cent) with fisheries as their main occupation (93 per cent).

Shinde *et al.* (1998) revealed that majority of the dairy farmers were in medium knowledge group followed by high and low knowledge group.

Meti (1998) reported majority of small and marginal farmers in Tungabhadra project area belonged to medium knowledge with respect to ground nut, cotton and paddy cultivation.

Gupta *et al.* (1999) studied on evaluation of adoption and impact of aquaculture into the farming systems in Bangladesh. The study revealed that about 49 per cent of farmers had education upto secondary or higher secondary level and 36 per cent had primary education and only 11.5 per cent were illiterate. Further, 54 per cent of the respondents taken farming as their principal occupation while for the rest it was the secondary occupation.

Awasthi *et al.* (2000) found that dairy farmers (Jabalpur district) had low to medium level of knowledge on improved dairy practices.

Manker *et al.* (2000) studied the socio economic and psychological profile of fishermen of Raigad district of Maharastra. The study indicated majority of fishermen had primary education (35 per cent), Joint family type (71.87 per cent), medium socio-economic status (73.33 per cent), medium income (70.00 per cent), medium risk orientation (80.00 per cent) and neutral attitude towards fishing occupation (60.00 per cent).

Talukdar (2000) found that majority (64 per cent) of fish farmers in Assam belonged to medium category followed by 23 percent and 13 percent in high and low categories, respectively. He observed that majority of fish farmers had medium level of cosmopoliteness and economic motivation. Further, he reported more than half of the fish farmers did not know the recommended dose of lime, manures/ inorganic fertilizers, suitable stocking density and recommended ratio of species to be stocked.

Singh and Nanda (2001) reported that majority of the fish farmers (72 per cent) in Madhya Pradesh belonged to scheduled tribes and other backward castes. 69 per cent of farmers had fishing as main occupation culturing exotic and Indian major carps. Three fourth of the farmers had seasonal ponds with rain as the only source of water. About 50 per cent of the fish farmers were operating small farms upto 1 ha. The annual income of about 70 per cent of farmers ranged from Rs. 15000 to Rs. 30000.

The study on socio-economic status of fishermen in relation to fisheries of Jaisamand Lake in Rajasthan has been made by Sharma and Anup Kumar (2001). This study highlighted high illiteracy, multi-occupation (more than one occupation) and livingness for nuclear family. Nearly 33 per cent of fishermen were landless while only 7 per cent of farmers had pucca houses whereas most of the farmers lived in kuchha houses. The authors have opined that slight income enhancement from fishing has not made any impact on improving of socio-economic status.

Mallick (2001) revealed that educational status of traditional fishers around Chilka Lake was limited to primary/middle school level and all the fishers are just above the poverty line.

Sushant *et al.* (2002) revealed that a majority of trapa-cum-fish farmers in Jabalpur district of Madhya Pradesh were middle aged, having low education status, large family size and belonged to medium annual income group with low extension contact and exposure to mass media where as their credit orientation was high.

Bandyopadhyay and Pradeep (2002) made detailed study on fisheries enhancement programmes in Oxbow lakes of West Bengal. The study revealed that majority of fishers were illiterate in kuchha houses. The income of fishermen families ranged between Rs. 11000 to Rs. 20000 and fishing equipments were possessed by only a few members (19 per cent).

Most of the fishers involved in different piscicultural activities in sewage fed water bodies in West Bengal belong to the backward communities (Ghosh *et al.*, 2002).

Mande, *et al.* (2008) reported that majority of Cattle owners have medium knowledge level about breeds of cattle, newly born calf, breeding, feeding, and Health management practices.

2.3 Attitude towards Agricultural and Allied Farm Enterprises

Pathak (1981) conducted the study in Barrakpore and Barasat in West Bengal and found out that farm size, tenure status, copping intensity, irrigation potentiality, aspirations, credit orientation, deffered gratification and empathy were significantly and positively correlated with attitude of farmers towards improved practices of jute cultivation.

Ponnappan (1982) found that 54 percent of FFDA beneficiaries in Madurai district of Tamil Nadu had favourable attitude towards FFDA while 27 and 19 percent of the respondents appeared to have most favorable and less favourable attitude respectively. Further, he also observed that 9.6 percent of the beneficiaries were satisfied with the functioning of FFDA programme.

Singh and Kunzru (1985) observed that 78.57 and 54.29 per cent of the goat and sheep rearing farmers had favourable attitude towards goat and sheep farming, while 2.86 and 18.57 per cent and 16.43 and 29.23 per cent of goat and sheep farmers had strongly favourable and unfavourable attitude, respectively.

Balasubramaniam (1988) revealed that majority of the fish farmers (91.25 per cent) possessed favourable attitude towards improved fish farming. The scale developed by him to measure attitude consisted of 10 items with 5 point continuum. Further, the author reported no significant relationship between adoption behavior and attitude towards improved fish farming practices in his study.

Navaneetha Krishnan (1988) reported variables like education, social participation, mass media exposure; extension agency contact, urban contact and socio-economic status were found to have influence on attitude level of both the members of FCS and non members in Madurai district (Tamil Nadu). Majority of the members and non members had neutral attitude towards Fisheries Cooperative Societies.

Srinath (1988) observed that owner fish farmers had better attitude towards scientific prawn farming than farmers, who had taken ponds on lease.

Haque (1989) stated that majority of fish farmers in West Bengal had favourable attitude towards composite fish culture.

Narayana Gowda (1992) reported that attitude towards watershed development programme had positive and significant relationship with overall consequence of watershed development programme in Karnataka.

Natarajan and Muthaiah (1995) observed that a large majority (73.33 per cent) of the farmers had favourable attitude while 17.5 per cent and 9.17 per cent of them had less favourable and more favourable attitude, respectively towards social forestry programme in Palani district (Tamil Nadu).

Prasad (1995) reported that majority of farmers in Ranga Reddy district of Andhra Pradesh belonged to less favourable attitude category towards dryland agricultural technologies followed by more favourable category.

Mahandra Kumar (1996) found that a larger percentage of FFDA farmers (68.6 per cent) had positive attitude towards fish farming whereas non FFDA fish farmers accounted for 21.9 per cent.

Subhas Chandra (1997) observed that 69 per cent of farmers in Andhra Pradesh had medium favourable attitude towards watershed management practices, while only 18 and 13 per cent had high favourable and less favourable attitude, respectively.

Trained beneficiaries of FFDA in Dharwad district (Karnataka) acquired favourable attitude on most aspects of fish culture as a result of training programme (Seenappa and Surendra, 1988). They measured attitude by incorporating six positive and four negative statements towards fish culture in the schedule and employed before and after design.

Meti (1998) majority of small and marginal farmers in TBP area had favourable attitude towards improved agricultural technologies.

Basavakumar and Yaligar (2000) reported that training helped in enhancement of attitude of trainees towards prawn cultivation in Uttara Kannada district of Karnataka.

Awasthi *et al.* (2000) observed that 50 per cent of dairy farmers had favourable attitude, followed by 35 per cent and 15 per cent of farmers had neutral and unfavourable attitude, respectively.

Govinda Gowda *et al.* (2000) found that large majority (65.5 per cent) of the dairy farmers possessed favourable attitude towards recommended dairy management practices whereas percentage of dairy farmers both in neutral and unfavourable attitude was equal.

Coastal fishermen of Orissa had favourable attitude towards the motorization of fishing carfts (Balasubramaniam *et al.*, 2001).

Panigrahi and Dana (2002) stated that traditional fish farmers of West Bengal showed least interest to adopt improved fish culture practices and it was found that they are not aware of scientific aquaculture.

Arpita Sharma (2002) in her studies on participation of women fishers in aquaculture activities reported a favourable attitude of women fishers towards adoption of aquaculture.

Sushant *et al.* (2002) in their studies on attitude of members of fishermen cooperative societies towards trapa-cum-fish farming found that the members stress more on trapa farming than fish farming as quick returns are obtained from it.

Nagarajaiah (2002) reported that 49 per cent of fish farmers of South Karnataka had more favourable attitude towards composite fish culture peactices, while 47 per cent had favourable attitude and remaining unfavourable.

Talukdar and Sontaki (2005) observed that 93.34 per cent of aquaculturists in Assam had favourable attitude towards composite fish culture practices, while only 1.66 per cent had more favourable attitude and remaining unfavourable. Further, he reported non-significant relationship between attitude of fish farmers toward composite fish culture practices and extent of adoption. The same relationship persisted even with knowledge level.

Thus, it can be inferred from the above review that there existed variation in the attitude of respondents towards the objects in question. Further, the above reviews also reveal that majority of respondents had a favourable attitude followed by moderate and less favourable attitude.

2.4 Association of Attitude with Independent Variables

Variable, Author and Year	*Area*	*Dependent Variable Attitude*	*Nature of Relationship*
Age			
Ponnappan (1982)	Tamil Nadu	Towards FFDA programme of fish farmers	Negatively non-significant
Navaneetha Krishnan (1988)	Madhurai, Tamil Nadu	Of members and non-members towards FCS	Negatively non-significant
Seenappa and Sundara (1988)	Dharwad, Karnataka	Of FFDA beneficiaries towards fish culture	Positively significant
Prasad (1995)	Ranga Reddy district, Andhra Pradesh	Towards dryland agricultural practices of farmers	Positively non-significant
Surekha *et al.* (1997)	Dharwad, Karnataka	Of farmers towards watershed development activities	Positively non-significant
Basavakumar and Yaligar (2000)	North Kanara, Karnataka	Of trainees towards prawn cultivation	Positively non-significant
Nijagonda (2000)	Bidar, Karnataka	Of red gram growers towards regulated markets	Positively non-significant
Family size			
Awasthi *et al.* (2000) association	Jabalpur	Towards improved dairy practices	Positively significant
Nijagonda (2000)	Bidar, Karnataka	Of redgram growers towards regulated markets	Positively non-significant
Nagarajaiah (2002)	Karnataka	Of composite fish culture practices	Negatively non-significant
Family type			
Dixit (1988)	Chitradurga district, Karnataka	Of farmers towards social and farm forestry	Positively significant
Nijagonda (2000)	Bidar, Karnataka	Of red gram growers towards regulated markets	Positively non-significant
Nagarajaiah (2002)	Karnataka	Of composite fish culture practices	Negatively non-significant

Contd...

Contd...

Variable, Author and Year	*Area*	*Dependent Variable Attitude*	*Nature of Relationship*
Education			
Ponnappan (1982)	Madhurai, Tamil Nadu	Towards FFDA programme	Negatively non-significant
Navaneetha Krishnan (1988)	Dharwad, Karnataka	Of non members towards FCS	Positively non-significant
Seenappa and Surendra (1988)	Dharwad, Karnataka	Of trainees towards fish culture	Negatively non-significant
Nimbal (1995)	Karnataka	Of farmers towards biogas technology	Positively non-significant
Mahandra Kumar (1996)	Tamil Nadu	Towards improved fish culture practices	Positively non-significant
Surekha *et al.* (1997)	Dharwad, Karnataka	Of farmers towards watershed development activities	Positively non-significant
Basavakumar and Yaligar (2000)	North Kanara, Karnataka	Of trainees towards prawn cultivation	Non significant
Awasthi *et al.* (2000)	Jabalpur	Towards improved dairy practices of dairy farmers	Positively non-significant association
Nijagonda (2000)	Bidar, Karnataka	Of red gram growers towards regulated markets	Positively non-significant
Roy and Pathak (2000)	Orissa	Of farm women towards post-harvest activities	Positively significant
Nagarajaiah (2002)	Karnataka	Of composite fish culture practices	Positively non-significant
Experience			
Ponnappan (1982)	Tamil Nadu	Towards FFDA programme	Negatively non-significant
Navaneetha Krishnan (1988)	Madhurai, Tamil Nadu	Of members and non-members towards FCS	Non-significant
Perumal *et al.* (1988)	Tamil Nadu	Of fish farmers towards FFDA	Positively non-significant
Biswas (1990)	Andaman and Nicobar Islands	Towards rice production technology	Positively significant
Raju Naik *et al.* (1994)	Dharwad, Karnataka	Towards seed supplying agencies	Non-significant
Prasad (1995)	Ranga Reddy district, Andhra Pradesh	Towards dryland agricultural technology	Positively non-significant

Contd...

Contd...

Variable, Author and Year	*Area*	*Dependent Variable Attitude*	*Nature of Relationship*
Shivashankarappa (1995)	Karnataka	Of participant of farmers towards watershed management	Positively non-significant
Awasthi *et al.* (2000)	Jabalpur	Towards improved dairy practices	Positively non-significant
Occupation status			
Navaneetha Krishnan (1988)	Madhurai, Tamil Nadu	Of members and non-members towards FCS	Non-significant
Nijagonda (2000)	Bidar, Karnataka	Of red gram growers towards regulated markets	Positively non-significant
Annual income			
Dixit (1988)	Karnataka	Of farmers towards social and farm forestry	Non-significant
Raju Naik *et al.* (1994)	Dharwad, Karnataka	Of farmers towards seed supplying agencies	Positively significant
Natarajan and Muthaiah (1995)	Tamil Nadu	Of farmers towards social forestry programme	Positively significant
Shivashankarappa (1995)	Karnataka	Of participant farmers towards watershed mgt.	Positively non-significant
Mahandra Kumar (1996)	Tamil Nadu	Towards improved fish culture practices	Positively significant
Nisha Aravind (1999)	Kerala	Of farmers towards homestead agro forestry	Positively significant
Awasthi *et al.* (2000)	Jabalpur	Towards improved dairy practices	Positively significant
Nijagonda (2000)	Bidar, Karnataka	Of red gram growers towards regulated markets	Positively non-significant
Roy and Pathak (2000)	Sundargarh, Orissa	Of farm women towards post harvest activities	Positively non-significant
Nagarajaiah (2002)	Karnataka	Of composite fish culture practices	Positively non-significant
Land holding			
Ponnappan (1982)	Madhurai, Tamil Nadu	Towards FFDA programme	Negative non-significant
Seenappa and Surendra (1988)	Dharwad, Karnataka	Towards fish culture by FFDA beneficiaries	Negatively non-significant
Dixit (1988)	Karnataka	Of farmers towards social & farm forestry programme	Positively non-significant
Raju Naik *et al.* (1994)	Karnataka	Of farmers towards seed supplying agencies	Positively significant
Nimbal (1995)	Dharwad, Karnataka	Of respondents towards biogas technology	Positively significant

Contd...

Contd...

Variable, Author and Year	*Area*	*Dependent Variable Attitude*	*Nature of Relationship*
Surekha *et al.* (1997)	Dharwad, Karnataka	Of farmers towards watershed development activities	Positively significant
Basavakumar and Yaligar (2000)	North Kanara, Karnataka	Of farmers towards prawn cultivation	Negatively significant
Roy and Pathak (2000)	Sundergarh, Orissa	Of farm women towards post harvest activities	Negatively significant
Nijagonda (2000)	Bidar, Karnataka	Of red gram growers towards regulated markets	Positively significant
Social participation			
Ponnappan (1982)	Madhurai district, Tamil Nadu	Towards FFDA programme	Non-significant
Dixit (1988)	Chitradurga district, Karnataka	Of farmers towards social and farm forestry	Non-significant
Prasad (1995)	Ranga Reddy district, Andhra Pradesh	Of farmers towards dryland agriculture practices	Positively significant
Mahandra Kumar (1996)	Tamil Nadu	Towards improved fish culture practices	Positively significant
Nisha Aravind (1999)	Kerala	Of farmers towards homestead agro-forestry	Positively significant
Nijagonda (2000)	Bidar, Karnataka	Of red gram growers towards regulated markets	Positively significant
Roy and Pathak (2000)	Sundergarh district, Orissa	Of farm women towards post harvest activities	Non-significant
Nagarajaiah (2002)	Karnataka	Of composite fish culture practices	Positively significant
Material possession			
Ponnappan (1982)	Madhurai district, Tamil Nadu	Towards FFDA programme	Negatively non-significant
Navaneetha Krishnan (1988)	Madhurai, Tamil Nadu	Of members and non-members towards FCS	Non-significant
Mass Media Participation			
Ponnappan (1982)	Madhurai district, Tamil Nadu	Towards FFDA programme	Negatively non-significant
Navaneetha Krishnan (1988)	Madhurai, Tamil Nadu	Of members and non-members towards FCS	Positively significant
Natarajan and Muthaiah (1995)	Palani district, Tamil Nadu	Of farmers towards social forestry programme	Positively significant
Prasad (1995)	Ranga Reddy district, Andhra Pradesh	Towards dryland agriculture practices	Positively significant

Contd...

Contd...

Variable, Author and Year	Area	Dependent Variable Attitude	Nature of Relationship
Surekha *et al.* (1997)	Dharwad, Karnataka	Of farmers towards watershed development activities	Non -significant
Nisha Aravind (1999)	Kerala	Of farmers towards homestead agro-forestry	Positively significant
Nijagonda (2000)	Bidar, Karnataka	Of redgram growers towards regulated markets	Positively non-significant
Nagarajaiah (2002)	Karnataka	Of composite fish culture practices	Positively non-significant
Extension agencies contact			
Ponnappan (1982)	Madhurai district, Tamil Nadu	Towards FFDA programme	Positively significant
Navaneetha Krishnan (1988)	Madhurai	Of non-members towards FCS	Positively significant
Natarajan and Muthaiah (1995)	Palani district, Tamil Nadu	Of farmers towards social forestry programme	Positively significant
Prasad (1995)	Ranga Reddy district, Andhra Pradesh	towards dryland agriculture practices	Positively significant
Awasthi *et al.* (2000)	Jabalpur	Towards improved dairy practices	Positively significant
Nagarajaiah (2002)	Karnataka	Of composite fish culture practices	Positively significant
Cosmopoliteness			
Dixit (1988)	Karnataka	Of farmers towards social forestry and farm forestry programme	Positively significant
Mahandra Kumar (1996)	Tamil Nadu	Towards improved fish culture practices	Positively significant
Surekha *et al.* (1997)	Karnataka	Of farmers towards watershed development	Positively significant
Awasthi *et al.* (2000)	Jabalpur	Towards improved dairy practices	Positively significant
Nagarajaiah (2002)	Karnataka	Of composite fish culture practices	Positively significant
Innovative proneness			
Ponnappan (1982)	Madhurai district, Tamil Nadu	Towards FFDA programme	Positively non-significant
Nagarajaiah (2002)	Karnataka	Of composite fish culture practices	Positively significant

Contd...

Contd...

Variable, Author and Year	*Area*	*Dependent Variable Attitude*	*Nature of Relationship*
Credit orientation			
Ponnappan (1982)	Madhurai, Tamil Nadu	Towards FFDA programme	Non-significant
Sunar Rao (1988)	Medak district, Andhra Pradesh	Towards improved dryland agricultural technology	Non-significant
Perumal *et al.* (1988)	Tamil Nadu	Of fish farmers towards FFDA	Positively non-significant
Shivashankarappa (1995)	North Karnataka	Of participant farmers towards watershed	Positively non-significant
Prasad (1995)	Ranga Reddy district, Andhra Pradesh	Of farmers towards dryland agricultural practices	Positively significant
Nagarajaiah (2002)	Karnataka	Of composite fish culture practices	Positively significant
Value orientation			
Dana(1987)	Uttar Pradesh	Towards artificial insemination in Cattle	Positively significant
Risk orientation			
Siddaramaiah and Reddy (1993)	Bangalore district, Karnataka	Of dryland farmers towards watershed recommended	Non-significant
Prasad (1995)	Ranga Reddy, Andhra Pradesh	Of farmers towards dryland agricultural practices	Positively significant
Mahandra Kumar (1996)	Tamil Nadu	Towards improved fish culture practice	Positively significant
Awasthi *et al.* (2000)	Jabalpur	Towards improved dairy practices	Positively significant
Balasubramaniam *et al.* (2001)	Orissa	Towards motorization of fishing crafts	Positively non-significant
Nagarajaiah (2002)	Karnataka	Of composite fish culture practices	Positively significant
Economic motivation			
Ponnappan (1982)	Madhurai, Tamil Nadu	Towards FFDA programme	Positively significant
Perumal *et al.* (1988)	Tamil Nadu	Of fish farmers towards FFDA	Positively significant
Prasad (1995)	Ranga Reddy district, Andhra Pradesh	Towards dryland agricultural technologies	Positively significant
Natarajan and Muthaiah (1995)	Palani district, Tamil Nadu	Of farmers towards social forestry programme	Positively significant

Contd...

Contd...

Variable, Author and Year	*Area*	*Dependent Variable Attitude*	*Nature of Relationship*
Mahandra Kumar (1996)	Tamil Nadu	Towards improved fish culture practices	Positively significant
Awasthi *et al.* (2000)	Jabalpur	Towards improved dairy practices	Positively significant
Nagarajaiah (2002)	Karnataka	Of composite fish culture practices	Positively significant
Knowledge			
Somasundram and Singh (1978)	–	Of small farmers regarding the high yielding varieties	Positively significant
Sidhu (1980)	Punjab	Towards dairy practices	Positively significant
Nagarajaiah (2002)	Karnataka	Of composite fish culture practices	Positively significant
Size of fish culture pond			
Ponnappan (1982)	Madhurai, Tamil Nadu	Towards FFDA programme	Positively significant
Mahandra Kumar (1996)	Tamil Nadu	Towards improved fish culture practice	Negatively significant
Nagarajaiah (2002)	Karnataka	Of composite fish culture practices	Non-significant
Duration of Water Availability			
Nagarajaiah (2002)	Karnataka	Of composite fish culture practices	Positively significant
Source of water			
Nagarajaiah (2002)	Karnataka	Of composite fish culture practices	Non-significant
Extent of weed infestation			
Nagarajaiah (2002)	Karnataka	Of composite fish culture practices	Non-significant

2.5 Adoption of Recommended Practices of Agricultural and Allied Farm Activities

Rogers and Shoemaker (1971) defined adoption as a decision to make full use of an innovation.

Ban and Hawkins (1988) defined adoption as a decision to apply to an innovation and to continuously use it.

Haque and Ray (1983a) reported 12 selected variable jointly explained 86 per cent variation in the behaviour of the fish farmers towards adoption of recommended species of fish in composite fish culture in West Bengal. The significant and extremely important variables were minimum depth of water in the pond, economic motivation, attitude towards composite fish culture and innovation proneness.

Meeran (1983) observed wide variation in the extent of overall adoption among fish farmers in Tamil Nadu. Though, majority of the fish farmers were found to have medium to high level of adoption, in general the extent of adoption of recommended practices was poor among the fish farmers. Practices like pond preparation and stocking of culturable species were adopted by a fairly high proportion of respondents. Nearly 1/3rd of the respondents adopted stocking density of fingerlings to the recommended level. The extent of adoption of other practices like application of lime, urea, super phosphate, muriate of potash was very low.

Nataraju and Channe Gowda (1984) reported that on the whole, 64 per cent of dairy farmers in Karnataka belonged to medium adoption category.

Balasubramaniam and Kaul (1985) observed that about 94 per cent of the marine fishermen fall in medium adoption category.

Subhash Chandra (1986) obseved that most of the fish farmers (83.33 per cent) in Andhra Pradesh were medium in their overall extent of adoption followed by high level of adoption (14.17 per cent). Only a limited percentage (2.50) had low extent of adoption. Among the practices studied fish pond preparation, application of lime, use of organic manures, stocking of culturable species, optimum stocking density and periodical netting were found to be accepted and adopted in varying degrees from low to very high by all the farmers, whereas varying degrees of non-adoption existed among practices like application of muriate of potash, feeding with proper feeds, application of inorganic fertilizers.

Reddy (1987) found that only 15 per cent of the respondents adopted the practice of feeding of dairy cattle with pre-mixed concentrate feed.

Chede (1988) found that only 13.51 per cent of dairy farmers had high level of adoption while more than 2/3rd had medium level of adoption of selected dairy technologies.

Das *et al.* (1988) reported that 55 per cent of variation in adoption of modern composite fish culture in West Bengal was attributed to 14 variables comprising situational, socio-personal, socio-psychological and communication factors.

Rajurkar (1988) studied the adoption of improved fish farming practices by farmers in Nagpur district of Maharashtra and reported that annual income, conservation and knowledge were positively related with extent of adoption of fish farming practices.

The findings of Balasubramaniam (1988) revealed majority of fish farmers were in medium adoption category while 15 per cent were in high adoption category and 13.75 per cent farmers exhibited low adoption behaviour. Out of 11 practices studied high partial adoption was observed with respect to control of aquatic weeds, application of manures and fertilizers, and supplementary feeding. Further, the author reported quite high adoption with respect to practices like stocking of recommended species of fish, correct stocking density and periodical netting to check the growth of fish.

Bose (1989) reported medium level of adoption of fish culture practices.

The study conducted by Krishnaiah (1989) in Andhra Pradesh revealed that majority of respondents were medium adopters while more than $1/4^{th}$ were high adopters and remaining were low adopters of improved fish culture practices. Among the different practices that were fully adopted by the fish farmers stocking density, supplementary feeding with rice bran and oil cake, and stocking of culturable species ranked first, followed by application of urea, manuring with cowdung, removal of predators and undesirable fish with subsequent ranks.

Wayal (1990) reported low adoption of improved fish farming practices among untrained fish farmers of Akola district.

The study conducted by Suresh *et al.* (1990) revealed that feed and fertilizers were applied far below the recommended levels which led to poor yields though stocking was done at the recommended rate among fish farmers of Tamil Nadu.

Ingle *et al.* (1991) reported low level of adoption although the knowledge was high. It was observed that 40.00 per cent of farmers belonged to low adoption category followed by 53.00 per cent of farmers in medium adoption category.

Santosh and Sudhir (1992) attributed low fish production under extensive systems in Ganjam district (Orissa) which was due to lack of proper management practices. It was observed that farmers were found to use inferior quality and low quality of different inputs such as manure, fertilizer, feed, seed, labour, etc. and stocking of seeds in the ponds was unscientific.

Study conducted by Bhaumik *et al.* (1992) in West Bengal revealed adoption of composite fish culture technology by 61.00 per cent of the fish farmers, out of which only 11.00 per cent adopted the package of practice as per recommendations for obtaining higher production and as many as 39.00 per cent of fish farmers surveyed did not adopt the modern composite fish culture technology. Further, they observed that farmers did not adopt all the recommended package of practices. However, control of predators and weed fishes was adopted cent per cent, whereas 49.07 per cent and 28.59 per cent of adopters applied manure fertilizer and supplementary feeds respectively. Practice of liming was adopted by majority of adopter farmers (97.52 per cent).

Praveena (1993) found that majority of fish farmers had low level of adoption of recommended fish culture practices.

Wayal and Ingle (1994) observed very poor adoption of fish culture practices except stocking with Indian major carps and common carp variety in Akola district (Maharashtra).

Majority of the farmers in Andhra Pradesh (Ranga Reddy district) had medium level of adoption of recommended cultivation practices of crops (Prasad, 1995).

Srinath (1996) observed very low of adoption of recommended practices particularly in areas where traditional prawn filtration.

Mahandra Kumar (1996) reported, adoption of composite fish practice in Tamil Nadu are inadequate. He found that nearly 2/3rd of FFDA fish farmers (69.00 per cent) in Tamil Nadu were adopting composite fish culture practices to medium extent while 37.14 per cent and 3.81 per cent of fish farmers, respectively categorized to high and low adoption group. Among the different practices studied, stocking density of fish seed (89.05 per cent), application of organic manure (43.33 per cent) and type of seed material used (41.00 per cent) were adopted by FFDA farmers. Further, he noticed low level of feeding and manuring practice by most of the fish farmers.

Reddy (1997) found that majority (70.65 per cent) of fish farmers in Nellore district belonged to medium adoption group. The percentage of farmers in low adoption and high adoption group were 19.33 per cent and 10.00 per cent respectively.

Mathiyalagan (1997) studied the acceptance pattern of poultry farm practices in Salem district of Tamil Nadu. Out of 14 practices studied 10 practices were adopted by almost all the respondents and other practices were adopted by certain percentage of respondents.

Bhaumik and Saha (1998) reported that 70 per cent of respondents in West Bengal were low adopters of composite fish culture while 22 percent and 8 percent belonged to medium and high adopter category.

Ramesh Kumar Reddy and Srivastava Rao (1998) in their study found that 52.33 per cent respondents had medium adoption level followed by 28.00 per cent with low adoption and 19.6 per cent with high adoption level of improved cabbage production technology.

Meti (1998) observed that majority of small and marginal farmers belonged to medium adoption categories with respect to improved cropping practices of groundnut, cotton and paddy. He opined that lack of knowledge, high cost and non-availability and unsuitability as major reasons for partial or non-adoption.

Mishra and Mishra (1999) found that only 10 per cent of the fish farmers in Orissa adopted the fish culture practice scientifically with the consultation of scientists of Krishi Vigyan Kendra (KVK) or Government officials. Further, they reported only 5 per cent of the farmers followed the integrated fish farming. The study also indicated that only 10 per cent of farmers used to put lime during pond preparation and management practices while manuring was done occasionally. Farmers rarely adopted use of inorganic fertilizers and supplementary feeding and production ranged from 200 to 700 kg/ha/annum.

Nagaraj (1999) found that a good number of big farmers (31.2 per cent) belonged to high adoption category while majority (37.5 per cent) of small farmers belonged to low level of adoption of groundnut production practices. However, when pooled sample was considered a majority of farmers (36.9 per cent) belonged to medium level of adoption category.

Meeran (2000) found that attributes namely education, experience, family size, material possession, extension agency contact, mass media exposure, scientific orientation influenced adoption of shrimp farming practices in Thanjavur and South Arcot districts of Tamil Nadu.

Venkatesh Prasad and Sidaramaiah (2000) reported that variables like risk orientation, family education, level of aspiration and innovation proneness have major direct effects on adoption of plant protection measures and also through which most of the other factors influenced the overall adoption of plant protection measures indirectly.

Meeran (2001) reported that majority of shrimp farmers (74.00 per cent) obtained yields in the range of 501-1500 kg/ha/crop in coastal district of Tamil Nadu under extensive systems and variables like experience, consultancy services availed, material possession and risk orientation were the most critical variables influencing the yield.

Panigrahi and Dana (2002) in their detailed work on Integrated VI Project in West Bengal revealed that 70.00 per cent of village ponds are under traditional culture practice and only 30.00 per cent village ponds were adopted with scientific culture method. Further, the authors reported production level of 1200-1400 kg/ha in improved fish culture practice against 300-400 kg/ha in traditional ponds.

Talukdar and Sontaki (2005) reported that majority of the aquaculturists (63.00 per cent) in Sonitpur district (Assam) had medium level of adoption of composite fish culture practices and the adoption behaviour of fish farmers was positively correlated with extension participation, economic motivation, cosmopoliteness and knowledge of fish farmers and negatively by their age.

Gautam *et al.* (2007) revealed that herd size, occupation, social participation, family education level were positively correlated with adoption of dairy practices, while farm size was highly negatively correlated with the adoption. The age, caste and family size were not significantly correlated with adoption of dairy practices.

Sarkar *et al.* (2007) revealed that the socio economic, socio psychological and communication factors have major direct effects on Rajbanshi Dairy farmers in relation to Animal Husbandry practices.

Thus, from the studies quoted above, it could be inferred that there was differential level of adoption of agricultural and composite fish culture practices by the farmers and fish farmers. In most of the studies referred above, farmers had medium level of adoption followed by low and high levels. However, satisfactory adoption was found with regard to certain practices of composite fish culture like pond preparation, selection of fish species, stocking density, periodical netting to check, growth of fish, while, it was not so in the case of practices like application of chemical fertilizers, feeding schedule, application of organic manures to enhance productivity of fish culture tanks and disease control measures.

2.6 Association between Adoption of Recommended Practices with Independent Variables

Variable, Author and Year	*Area*	*Dependent Variable Adoption*	*Nature of Relationship*
Age			
Meeran (1983)	Tamil Nadu	Of improved fish culture practices	Positively non-significant
Balasubramaniam and Kaul (1985)	Kerala coast	Of improved practices by traditional fishermen	Non-significant
Subhash Chandra (1986)	Andhra Pradesh	Of composite fish culture practices by fish farmers	Positively non-significant
Balasubramaniam (1988)	Tamil Nadu	Of improved fish culture practices by fish farmers	Negatively non-significant
Sujath Kumar (1988)	Tamil Nadu	Of improved fishing technologies by traditional fishermen	Negatively significant
Das *et al.* (1988)	West Bengal	Of composite fish culture	Negatively significant
Krishnaiah (1989)	Andhra Pradesh	of composite fish culture practices	Positively significant
Bhaumik *et al.* (1992)	West Bengal	Of composite fish culture	Positively significant
Brajmohan *et al.* (1993)	Ernakulam	Technological gap score	Non-significant
Mathiyalagan and Subramanian (1995)	Salem, Tamil Nadu	Of improved poultry practices	Negatively non-significant
Reddy (1997)	Andhra Pradesh	Level of fish farmers	Positively significant
Nagaraj *et al.* (2000)	Tumkur, Karnataka	Of groundnut production technology	Positively significant
Meeran (2000)	Tamil Nadu	Of shrimp farming	Negatively non-significant
Nagarajaiah (2002)	Karnataka	Of composite fish culture practices	Negatively non-significant
Talukdar and Sontaki (2005)	Sonitpur, Assam	Of composite fish culture practices	Negatively significant
Gautam *et al.* (2007)	–	Of dairy practices	Non-significant

Contd...

Contd...

Variable, Author and Year	*Area*	*Dependent Variable Adoption*	*Nature of Relationship*
Caste			
Subhash Chandra (1986)	Kurnool, Andhra Pradesh	Of composite fish culture practices	Negatively non-significant
Bhaumik *et al.* (1992)	West Bengal	Of composite fish culture technology	Positively significant
Reddy (1997)	Nellore, Andhra Pradesh	Level of fish farmers	Positively significant
Nagarajaiah (2002)	Karnataka	Of composite fish culture practices	Positively non-significant
Gautam *et al.* (2007)	–	of dairy practices	Non-significant
Family size			
Balasubramaniam (1988)	Tamil Nadu	Of improved fish culture practices	Positively non-significant
Reddy (1997)	Andhra Pradesh	Level of fish farmers	Negatively non-significant
Sah and Ram Chand (1999)	Banka, Bihar	Of improved dairy farming practices	Non-significant
Nagaraj *et al.* (2000)	Tumkur, Karnataka	Of groundnut technology	Positively non-significant
Nagarajaiah (2002)	Karnataka	Of composite fish culture practices	Positively non-significant
Gautam *et al.* (2007)	–	of dairy practices	Non-significant
Family type			
Mathiyalagan and Subramanian (1995)	Salem, Tamil Nadu	Of improved poultry practices	Positively non-significant
Reddy (1997)	Andhra Pradesh	Level of fish farmers	Negatively non-significant
Mathiyalagan (1997)	Tamil Nadu	Level of poultry practices	Positively non-significant
Nagarajaiah (2002)	Karnataka	Of composite fish culture practices	Negatively non-significant
Education			
Meeran (1983)	Tamil Nadu	Of improved fish culture practices	Positively significant
Subhash Chandra (1986)	Andhra Pradesh	Of composite fish culture practices	Negatively non-significant
Balasubramaniam (1988)	Tamil Nadu	Of improved fish culture practices	Positively non-significant

Contd...

Contd...

Variable, Author and Year	*Area*	*Dependent Variable Adoption*	*Nature of Relationship*
Sujath Kumar (1988)	Tamil Nadu	Of improved fishing technology by traditional fishermen	Negatively non-significant
Das *et al.* (1988)	West Bengal	Of composite fish culture practices	Positively significant
Sheoran and Ram Kumar (1988)	Karnal, Haryana	Of dairy innovations	Positively non-significant
Krishnaiah (1989)	Andhra Pradesh	Of composite fish culture practices	Positively significant
Ingle *et al.* (1991)	Vidarbha	Of improved fish farming practices by members of FCS	Positively non-significant
Bhaumik *et al.* (1992)	West Bengal	Of composite fish culture practices	Positively significant
Mathiyalagan and Subramanian (1995)	Tamil Nadu	Of improved poultry practices	Positively significant
Mahandra Kumar (1996)	Tamil Nadu	Of composite fish culture practices	Positively significant
Reddy (1997)	Andhra Pradesh	Level of fish farmers	Positively significant
Shinde *et al.* (1999)	Nagpur, Maharashtra	Of improved animal husbandry practices	Positively significant
Nagaraj *et al.* (2000)	Tumkur, Karnataka	Of groundnut technology	Positively significant
Nagarajaiah (2002)	Karnataka	Of composite fish culture practices	Non-significant
Talukdar and Sontaki (2005)	Sonitpur, Assam	Of composite fish culture practices	Non-significant association
Gautam *et al.* (2007)	–	of dairy practices	Positively significant
Experience			
Subash Chandra (1986)	Kurnool district, Andhra Pradesh	Of composite fish culture practices	Positively non-significant
Balasubramaniam (1988)	Fish farmers, Tamil Nadu	Of improved fish culture practices	Negatively non-significant
Sujath Kumar (1988)	Traditional fishermen, Tamil Nadu	Of improved fishing technologies	Negatively non-significant
Das *et al.* (1988)	West Bengal	Of composite fish culture practices	Positively significant
Krishnaiah (1989)	Andhra Pradesh	Of composite fish culture practices	Positively significant

Contd...

Contd...

Variable, Author and Year	*Area*	*Dependent Variable Adoption*	*Nature of Relationship*
Braj Mohan *et al.* (1993)	Ernakulam	Technological gap score	Positively significant
Mathiyalagan and Subramanian (1995)	Tamil Nadu	Of improved poultry practices	Positively non-significant
Reddy (1997)	Andhra Pradesh	Level of fish farmers	Positively significant
Mathiyalagan (1997)	Tamil Nadu	Of poultry practices	Positively non-significant
Shinde *et al.* (1999)	Nagpur	Of improved animal husbandry practices	Negatively non-significant
Meeran (2000)	Tamil Nadu	Level of shrimp farming	Positively significant
Subashini and Thyagarajan (2000b)	S. A. Vallalar, Tamil Nadu	Of tapioca technologies	Positively non-significant
Nagarajaiah (2002)	Karnataka	Of composite fish culture practices	Positively non-significant
Chandrakala and Eswarappa (2001)	Bangalore, Karnataka	Of improved dairy management practices	Positively significant
Occupation			
Meeran (1983)	Tanjavur, Tamil Nadu	Of fish culture practices	Non-significant
Subhash Chandra (1986)	Kurnool, Andhra Pradesh	Of composite fish culture practices	Negatively non-significant
Krishnaiah (1989)	Andhra Pradesh	Of composite fish culture practices	Positively non-significant
Mathiyalagan and Subramanian (1995)	Salem, Tamil Nadu	Of improved poultry practices	Positively non-significant
Mahandra Kumar (1996)	Tamil Nadu	Of composite fish culture practices	Positively significant
Mathiyalagan (1997)	Tamil Nadu	Of poultry practices	Positively non-significant
Meeran (2000)	Tamil Nadu	Level of shrimp farming	Positively non-significant
Nagarajaiah (2002)	Karnataka	Of composite fish culture practices	Positively significant
Gautam *et al.* (2007)	–	Of dairy practices	Positively significant

Contd...

Contd...

Variable, Author and Year	*Area*	*Dependent Variable Adoption*	*Nature of Relationship*
Annual Income			
Subash Chandra (1986)	Kurnool, Andhra Pradesh	Of composite fish culture practices	Negatively non-significant
Balasubramaniam (1988)	Tamil Nadu	Of improved fish culture practices	Positively non-significant
Chede (1988)	Maharashtra	Of selected dairy technologies	Positively significant
Sujath Kumar (1988)	Traditional fishermen, Tamil Nadu	Of improved fishing technologies	Positively significant
Mathiyalagan and Subramanian (1995)	Salem district, Tamil Nadu	Of improved poultry practices	Positively significant
Mahandra Kumar (1996)	Tamil Nadu	Of composite fish culture practices	Positively significant
Mathiyalagan (1997)	Tamil Nadu	Of poultry practices	Positively significant
Shinde *et al.* (1999)	Nagpur	Of improved animal husbandry practices	Positively significant
Sah and Ram Chand (1999)	Banka, Bihar	Of improved dairy farming practices by farmers	Positively significant
Nagarajaiah (2002)	Karnataka	Of composite fish culture practices	Positively significant
Talukdar and Sontaki (2005)	Assam	Of composite fish culture practices	Positively non-significant
Land holding/farm size			
Meeran (1983)	Tamil Nadu	Of improved fish culture practices	Non-significant
Sheoran and Ram Kumar (1988)	Karnal, Haryana	Of dairy innovations	Positively significant
Chede (1988)	Maharashtra	Of selected dairy technologies	Positively significant
Ingle *et al.* (1991)	Vidarbha	Of improved fish farming practices by members of FCS	Positively non-significant
Mathiyalagan and Subramanian (1995)	Salem, Tamil Nadu	Of improved poultry practices	Positively non-significant
Reddy (1997)	Andhra Pradesh	Level of fish farmers	Positively significant
Mathiyalagan (1997)	Tamil Nadu	Of poultry practices	Positively non-significant

Contd...

Contd...

Variable, Author and Year	*Area*	*Dependent Variable Adoption*	*Nature of Relationship*
Shinde *et al.* (1999)	Nagpur, Maharashtra	Of improved animal husbandry practices	Positively significant
Sah and Ram Chand (1999)	Banka, Bihar	Of improved dairy farming practices by farmers	Positively significant
Shinde *et al.* (1999)	Maharashtra	Of improved animal husbandry practices	Positively significant
Nagaraj *et al.* (2000)	Tumkur, Karnataka	Of groundnut technology	Positively significant
Venkatesh Prasad and Siddaramaiah(2000)	Karnataka	Of plant protection measures	Positively significant
Nagarajaiah (2002)	Karnataka	Of composite fish culture practices	Positively non-significant
Talukdar and Sontaki (2005)	Assam	Of composite fish culture practices	Positively non-significant
Social participation			
Meeran (1983)	Tamil Nadu	Of improved fish culture practices	Positively significant
Subhash Chandra (1986)	Kurnool, Andhra Pradesh	Of composite fish culture practices	Positively non-significant
Sheoran and Ram Kumar (1988)	Karnal, Haryana	Of dairy innovations	Positively non-significant
Balasubramaniam (1988)	Tamil Nadu	Of improved fish culture practices	Positively non-significant
Sujath Kumar (1988)	Tamil Nadu	Of improved fishing technologies by traditional fishermen	Positively significant
Krishnaiah (1989)	Andhra Pradesh	Of recommended fish culture practices	Positively significant
Ingle *et al.* (1991)	Vidarbha	Of improved fish farming practices by members of FCS	Positively non-significant
Mahandra Kumar (1996)	Tamil Nadu	Of composite fish culture practices	Positively significant
Reddy (1997)	Nellore, Andhra Pradesh	Level of fish farmers	Positively significant
Nagaraj *et al.* (2000)	Tumkur, Karnataka	Of groundnut technology	Positively significant
Meeran (2000)	Tamil Nadu	Level of shrimp farming	Positively non-significant
Venkatesh Prasad and Siddaramaiah (2000)	Karnataka	Of plant protection measures	Positively significant

Contd...

Contd...

Variable, Author and Year	*Area*	*Dependent Variable Adoption*	*Nature of Relationship*
Nagarajaiah (2002)	Karnataka	Of composite fish culture practices	Positively significant
Talukdar and Sontaki (2005)	Sonitpur, Assam	Of composite fish culture practices	Negatively non-significant
Gautam *et al.* (2007)	–	of dairy practices	Positively significant
Material possession			
Meeran (1983)	Tamil Nadu	Of fish culture practices	Significant association
Subhash Chandra (1986)	Kurnool, Andhra Pradesh	Of composite fish culture practices	Positively significant
Sujath Kumar (1988)	Tamil Nadu	Of improved fishing technologies by trawler owners	Positively significant
Krishnaiah (1989)	Andhra Pradesh	Of recommended fish culture practices	Positively significant
Mathiyalagan and Subramanian (1995)	Tamil Nadu	Of improved poultry practices	Positively non-significant
Mathiyalagan (1997)	Tamil Nadu	Of poultry practices	Positively non-significant
Meeran (2000)	Tamil Nadu	Level of shrimp farming	Positively significant
Mass media participation			
Balasubramaniam (1988)	Tamil Nadu	Of improved fish culture practices	Positively non-significant
Sujath Kumar (1988)	Tamil Nadu	Of improved fishing technologies by fishermen	Positively significant
Das *et al.* (1988)	West Bengal	Of composite fish culture practices	Positively significant
Sheoran and Ram Kumar (1988)	Karnal, Haryana	Of dairy innovations	Positively non-significant
Mathiyalagan (1997)	Tamil Nadu	Of poultry practices	Positively significant
Nagaraj *et al.* (2000)	Tumkur, Karnataka	Of groundnut production technology	Positively significant
Meeran (2000)	Tamil Nadu	Level of shrimp farming	Positively significant
Venkatesh Prasad and Siddaramaiah (2000)	Karnataka	Of plant protection measures	Positively non-significant

Contd...

Contd...

Variable, Author and Year	*Area*	*Dependent Variable Adoption*	*Nature of Relationship*
Subhashini and Thyagarajan (2000b)	S.A. Vallalar, Tamil Nadu	Of tapioca technologies	Positively significant
Nagarajaiah (2002)	Karnataka	Of composite fish culture practices	Positively significant
Extension Agency contact			
Meeran (1983)	Tamil Nadu	Of fish culture practices	Positively significant
Subhash Chandra (1986)	Kurnool, Andhra Pradesh	Of composite fish culture practices	Positively significant
Balasubramaniam (1988)	Tamil Nadu	Of improved fish culture practices	Positively significant
Sheoran and Ram Kumar (1988)	Karnal, Haryana	Of dairy innovations	Negatively significant
Bhaumik *et al.* (1992)	West Bengal	Of composite fish culture technology	Positively significant
Mathiyalagan (1997)	Tamil Nadu	Of poultry practices	Positively non-significant
Shinde *et al.* (1999)	Nagpur, Maharashtra	Of improved animal husbandry practices	Positively significant
Nagaraj *et al.* (2000)	Tumkur, Karnataka	Of groundnut technology	Positively significant
Meran (2000)	Tamil Nadu	Level of shrimp farming	Positively significant
Subashini and Thyagarajan (2000b)	S.A. Vallalar, Tamil Nadu	Of tapioca technologies	Positively significant
Chandrakala and Eswarappa (2001)	Bangalore, Karnataka	Of improved dairy management practices	Positively significant association
Nagarajaiah (2002)	Karnataka	Of composite fish culture practices	Positively significant
Talukdar and Sontaki (2005)	Sonitpur, Assam	Of composite fish culture practices	Non- significant association
Cosmopoliteness			
Mahandra Kumar (1996)	Tamil Nadu	Of composite fish culture practices	Positively significant
Reddy (1997)	Nellore, Andhra Pradesh	Level of fish farmers	Positively significant
Shinde *et al.* (1999)	Nagpur, Maharashtra	Of improved animal husbandry practices	Positively significant
Nagaraj *et al.* (2000)	Tumkur, Karnataka	Of groundnut technology	Positively non-significant
Venkatesh Prasad and Siddaramaiah (2000)	Karnataka	Of plant protection measures	Positively significant

Contd...

Contd...

Variable, Author and Year	*Area*	*Dependent Variable Adoption*	*Nature of Relationship*
Chandrakala and Eswarappa (2001)	Bangalore, Karnataka	Of improved fish management practices	Positively significant
Nagarajaiah (2002)	Karnataka	Of composite fish culture practices	Positively significant
Talukdar and Sontaki (2005)	Assam	Of composite fish culture practices	Positively significant
Innovative proneness			
Moulik (1965)	Delhi territory	Individual's degree of self-rating	Positively significant
Biswas *et al.* (1991)	West Bengal	Of composite fish culture practices	Positively significant
Nagarajaiah (2002)	Karnataka	Of composite fish culture practices	Positively significant
Credit orientation			
Sujath Kumar (1988)	Tamil Nadu	Of improved fishing technologies by traditional fishermen	Positively non-significant
Mathiyalagan and Subramanian (1995)	Salem, Tamil Nadu	Of improved poultry practices	Positively non-significant
Mathiyalagan (1997)	Salem, Tamil Nadu	Of poultry practices	Positively non-significant
Nagaraj (1999)	Chitradurga, Karnataka	Level of groundnut growers	Non-significant
Nagarajaiah (2002)	Karnataka	Of composite fish culture practices	Positively significant
Risk orientation			
Chede (1988)	Maharashtra	Of selected dairy technologies	Non- significant
Biswas *et al.* (1991)	West Bengal	Of composite fish culture practices	Positively significant
Reddy (1997)	Nellore, Andhra Pradesh	Level of fish farmers	Negatively significant
Krishnamurthy (1997)	Chitradurga, Karnataka	Of improved dairy practices	Positively significant
Nagaraj (1999)	Tumkur, Karnataka	Level of groundnut growers	Positively significant
Meeran (2000)	Tamil Nadu	Of shrimp farming	Positively significant

Contd...

Contd...

Variable, Author and Year	*Area*	*Dependent Variable Adoption*	*Nature of Relationship*
Venkatesh Prasad and Siddaramaiah (2000)	Karnataka	Of plant protection measures	Positively significant
Subhashini and Thyagarajan (2000b)	S.A. Vallalar, Tamil Nadu	Of tapioca technologies	Positively significant
Nagarajaiah (2002)	Karnataka	Of composite fish culture practices	Positively significant
Talukdar and Sontaki (2005)	Sonitpur, Assam	Of composite fish culture practices	Positively significant
Economic motivation			
Biswas *et al.* (1991)	West Bengal	Of composite fish culture practices	Positively significant
Nagarajaiah (2002)	Karnataka	Of composite fish culture practices	Positively significant
Talukdar and Sontaki (2005)	Sonitpur, Assam	Of composite fish culture practices	Positively significant
Knowledge			
Talukdar and Sontaki (2005)	Sonitpur, Assam	Of composite fish culture practices	Positively significant
Size of water body			
Subhash Chandra (1986)	Karnool, Andhra Pradesh	Of composite fish culture practices	Positively non-significant
Misra (1987)	West Bengal	Of composite fish farming	Negatively significant
Balasubramaniam (1988)	Tamil Nadu	Of improved fish culture practices	Negatively non-significant
Das *et al.* (1988)	West Bengal	Of composite fish culture	Negatively significant
Krishnaiah (1989)	Andhra Pradesh	Of recommended fish culture practices	Positively significant
Mahandra Kumar (1996)	Tamil Nadu	Of composite fish culture practices	Negatively significant
Talukdar (2000)	Assam	Of composite fish culture practices	Positively non-significant
Nagarajaiah (2002)	Karnataka	Of composite fish culture practices	Positively non-significant
Meeran (2000)	Tamil Nadu	Level of shrimp farming	Positively significant
Talukdar and Sontaki (2005)	Sonitpur, Assam	Of composite fish culture practices	Non- significant

Contd...

Contd...

Variable, Author and Year	*Area*	*Dependent Variable Adoption*	*Nature of Relationship*
Duration of water availability			
Balasubramaniam (1988)	Tamil Nadu	Of improved fish culture practices	Negatively non-significant
Nagarajaiah (2002)	Karnataka	Of composite fish culture practices	Positively non-significant
Source of water			
Balasubramaniam (1988)	Tamil Nadu	Of improved fish culture practices	Positively non-significant
Nagarajaiah (2002)	Karnataka	Of composite fish culture practices	Negatively non-significant
Extent of weed infestation			
Nagarajaiah (2002)	Karnataka	Of composite fish culture practices	Positively significant

2.7 Constraints to Adoption of Scientific Fish Culture

Analysis of constraints faced by the farmers in the adoption of recommended practices helps to identify the factors (other than identified independent variables) responsible directly or indirectly for poor adoption or non-adoption or partial adoption. This kind of study helps to identify problems like local problems, situational problems, organizational, social and technical problems which significantly affect the extent of adoption of fish culture practices.

Ponnappan (1982) observed that most of the beneficiaries under FFDA progamme expressed non-distribution of fingerlings at the tank spot as the main problem followed by other major problems like high mortality of fingerlings, low price for the fish produced, inadequate credit to met the cost of inputs like feed, fertilizer etc., inadequate field extension staff and lack of market conveyance.

Haque and Ray (1983b) reported ten problems in composite fish culture as perceived by fish farmers of West Bengal, out of which scarcity of fish seed of exotic carps, lack of information about the technology, lack of pure seed of major carps and non-availability of credit for fish culture occupied top four problems.

Balasubramaniam and Kaul (1985) revealed financial (63.75 per cent) constraints and unawareness of technology (36.25 per cent) as the dominant problems in the use of improved practices among traditional fishermen of Kerala.

Srinath (1988) reported lack of detailed knowledge about the technology, paucity of seed, inability to draw periodical income and risk involved due to leased holding as major constraints in shrimp farming.

Bhaumik *et al.* (1988) revealed that non-availability of credit, exotic carp seed, lack of knowledge, poaching, lesser consumer preference, poisoning of fish pond were the constraints faced in adoption of improved practices of fish culture.

Chede (1988) reported lack of knowledge on scientific animal feeding, milk preservation practice and scientific animal management as major constraints in adoption of dairy technologies in addition to non-availability of loan facilities for purchase of animals.

Balasubramaniam (1988) observed lack of timely and adequate supply of fish seed, non-availability of water for fish culture, supply of fingerling to the pond site and lack of finance as important constraints limiting the adoption of improved fish culture practices in Tamil Nadu. Constraints like poaching, presence of predator and weed fishes, multi use of water bodies and inadequate extension support from the officials were also felt by the fish farmers for better adoption.

Krishniah (1989) reported six constraints namely water changing, water level maintenance, lack of transport and storage facilities for fish, untimely credit facilities, price fluctuation for fish and non-availability of inputs within farmers' reach are found to be important problems encountered by fish farmers in adoption of recommended practices in Andhra Pradesh.

Bhaumik *et al.* (1992) stated non-availability of finance followed by inadequate security, poaching and poisoning were the major constraints for non-adoption of the composite fish culture technology in West Bengal. The other factors like lack of knowledge, non-availability of ideal water bodies and multiple ownership of the ponds were also the reasons for non-adoption.

Pounraj (1992) reported that inland fishermen of fisheries cooperative societies in Tanjavur district of Tamil Nadu expressed lack of adequate credit supply (87.50 per cent), inadequate supply of fish farming inputs (84.16 per cent), insufficient amount of subsidy (80.00) and inadequate knowledge about integrated fish farming (72.50 per cent) as the major constraints in fish farming enterprise. A few members also expressed poaching and complete failure of fish culture by disease attack as constraints in fish farming.

Wayal and Ingle (1994) found that loss of fish seed stocked through outlets in rainy season (63.26 per cent) and mortality due to predators (58.16 per cent) were the major production constraints and lack of training facilities and inadequate technical guidance were the extension constraints expressed by fishermen of Akola district.

Taufiq Rusdi (1994) found that fluctuation in market price (89 per cent), high cost of input (85 per cent) and delayed cash payment (72 per cent) are the dominant problems faced by ornamental fish farmers in Jakarta.

Beerannavar (1995) in his study on dairy management practices by trained farmers observed that inadequate own capital (76.74 per cent), non-availability of loan (41.88 per cent), lack of milk marketing facilities (30.23 per cent) as prime three reasons for not undertaking dairy enterprise.

Meeran and Jayaseelan (1996), while working on problems perceived by shrimp farmers reported that high fluctuations of water, poor quality seed, high cost of seed and feeds, inadequate extension services, lack of laboratory facilities and government support were the major problems.

Srinath (1996) reported high cost of farming, lack of technical knowledge, finance, risk due to mortality as dominant constraints related to adoption of shrimp farming in Kerala.

Alam (1997) reported that financial constraints, lack of fish seed and transportation facilities, short leasing period, organized marketing, etc. as the major problems faced by FFDA farmers in Assam.

Bhaumik and Saha (1998) found that high input cost, poaching of fish, poisoning of fish pond, high rent for water bodies, lack of follow-up activities by extension staff, non-availability of finance, multiple ownership of water body, not obtaining the expected growth of fish, etc. as the problems in fish farming.

Mishra and Mishra (1999) observed that lack of knowledge about the effect of fertilizers and manure in the pond and non-availability of inputs etc., as main reason for non-adoption or partial adoption in Orissa.

The study conducted by Talukdar (2000) found nearly half of the respondents (48 per cent) faced financial problem to practice composite fish culture followed by 43 per cent who faced problem of non-availability of good quality fish seed. About 38 per cent of farmers had problem of growth of fish not upto the expected level thereby decrease in production level. Poaching and poisoning of pond were not significant problems in Assam.

Nayak *et al.* (2001) observed that problem of white spot disease, lack of credit facilities and delayed resolution of legal problems as major constraints encountered by the shrimp farmers in Orissa.

Sharma (2002) reported that problems of women fishers in different geographical locates might be different according to local situations which need to be immediately recognized and provide support. The major constraints identified were related to finance, knowledge, expertise, socio-culture and infrastructure.

Chapter 3

Research Setting

In a social science research of this kind, a proper selection and interpretation of the concerned data presupposes a fair understanding of the locale and the people under study. The researcher must be familiar with behaviour complex of the inhabitants of the area under reference. An understanding of the socio-demographic background of the respondents is also necessary.

There are two types of research settings, viz, those at the macro and micro level respectively. Macro level encompasses the state as a whole while micro level ranges from the boundaries of the selected district to the block and a village periphery. The rationale of this kind of presentation is the imperative need to internalize the study environment from two perspectives, one for an extensive outlook and other for an intensive one. Firstly, the locale is to be viewed in terms of the boarder perspective with the state as the reference frame from and then it is considered at the more restricted level of the district and block as the unit for the in-depth study.

Macro Research Setting

Under macro research setting general information about india and West Bengal has been incorporated.

India

Main land of India extends in the tropical and subtropical zones from latitude $8^{0}4'$ to $37^{0}6'$ North and from longitude $68^{0}7'$ to $97^{0}25'$ East. The southern most point in India territory, the indira point (finally called Pygmalion point) is situated at $6^{0}30'$ North in the Nicobar Island. The country thus lies wholly in the northern and eastern hemispheres. The northern most point of India lies in the states of Jammu and Kashmir and it is known as Indira Col.

Area and Boundaries

India stretches 3214 km at its maximum from north to south and 2933 km at its

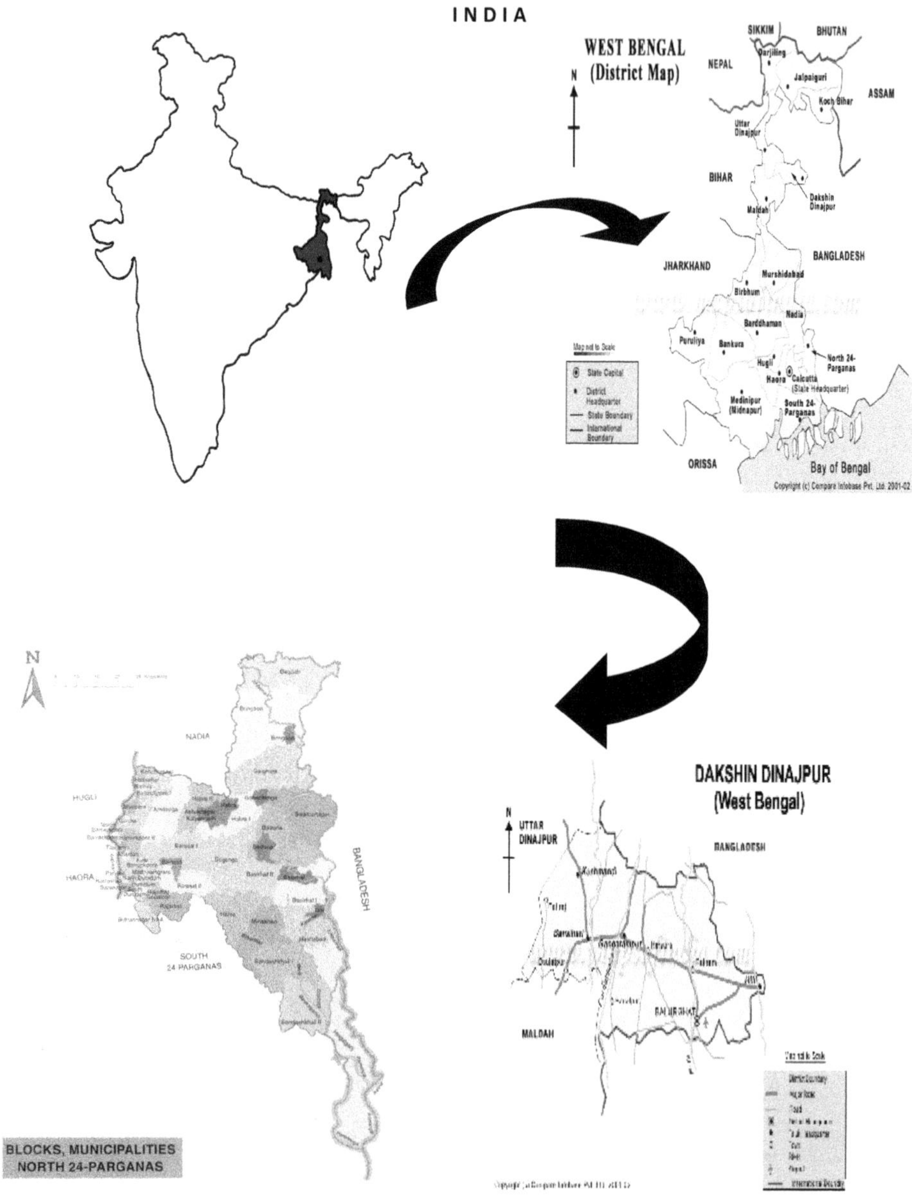

Figure 2: Map of India, West Bengal, North 24 Parganas and Dakshin Dinajpur District.

maximum from east to west. The total length of the main land coastal line is nearly 6100 km and the land frontier measures 15200 km.

The total length of the coastal line including that of the Island is about 7500 km with an area of 32,87,782 sq. km. India is the seventh largest country in the world, accounting for about 2.4 per cent of total world area. In terms of population, India is second only to China accounts for 16.87 per cent of the total world population.

India's neighbour in the north are China, Nepal and Bhutan. To the north east, India shares a boundary mainly with Pakistan and to the east with Myanmar, while Bangladesh forms almost an enclave within India. Afghanistan is another close neighbour of India towards north-west. The country is shaped somewhat line a triangle with its base in the north (the Himalayas) and the narrow apex in the south (Kanyakumari). South of the topic of cancer, the India mass tapers between the Bay of Bengal in the east and Arabian Sea in the west, the Indian ocean lies south. The Bay of Bengal and the Arabian Sea are its two northward extension. In the south on the eastern side the gulf of Mannar and the palk straight separate India from Sri Lanka, Indian's Islands includes Andaman Nicobar Islands in the Bay of Bengal and Lacadive (Lakshadweep), Minicoy and Amindive Island in the Arabian sea.

State of West Bengal

Out of 28 states of India, West Bengal stands 13th in area and ranks fourth in population wise. The state is located at 21^{0}38′ to 27^{0}10′ north latitude and 85^{0}50′ to 89^{0}5′ E longitude. Nepal in the north, Assam in the north east. West Bengal is bounded by Bangladesh in the east, Bay of Bengal in the south, Orissa in the south-east, Jharkhand in the north-west.

Basic Information

- ☆ **Area** : 88752 Sq. Km
- ☆ **Population** : 91276115
- ☆ **Sex ratio** : 950
- ☆ **Density/Sq Km.** : 1028
- ☆ **Literacy rate** : 76.26 per cent
- ☆ **Capital** : Kolkata
- ☆ **No.of districts** : 18
- ☆ **Assembly seats** : 294
- ☆ **Loksabha seats** : 42
- ☆ **Rajya sabha seats** : 16

Micro Research Setting

The basic information about the district is discussed in the section.

North 24-Parganas

Location and Boundary

North 24-Parganas district lies between $23^{0}15'2''$ and $22^{0}11'6''$North latitude and $89^{0}5'$ and $88^{0}20'$ East longitudes. Its territory extends over an area of 2219 sq. kms. The district is bounded by Nadia district on the north, South 24 Parganas district on the South. Hoogly river is in western and Bangladesh is in the eastern side of the district. The district comprises of twenty two blocks.

Climate

The district is situated in the sub-tropical climate region. Mainly three seasons that is summer, rainy (monsoon) and winter have been observed. The average annual rainfall is 1,579 mm (normal) with about 66 rainy days. In the summer season, the maximum temperature is 41^{0}C. In winter season, the minimum temperature is 10ºC. The maximum rainfall occurs during July month. Relative Humidity is 50 per cent in March and 90 per cent in July.

Soil Status

It is situated in new alluvium sub-region of the lower Gangetic plain (Zone lll). The type of soil varies from sandy to clayey loam, sandy loam being predominant. Ratio of land is High : Medium : Low=17 : 44 : 39. The land in this district falls in the most fertile tract of the state with neutral soil.

Area and Population

The district of North 24-Parganas was inhabited by 10,009,781 persons of whom 5119389 (51.14 per cent) were men and 4890392 (48.86 per cent) women, the sex ratio worked out to 955 women per 1000 men as per 2011 census. The urban population of the district as per 2011 census was 5732162 (57.27 per cent of the total district population) distributed among the two towns of the district. The rural population of the district was 4277619 (42.73 per cent of the total district population), according to the 2011 census, distributed among 1,518 inhabited villages.

Working Population

The total workers of the district were 3,571,624 and accounted for about 35.68 per cent of the total population of the district. The number of population were as cultivators which accounted about 288,058 (8.07 per cent of total workers). As per 2011 census, the total Agricultural Labourers was 599,039 which accounted about 16.77 per cent of total workers).

Agriculture

The district produces different types of food grains round the year. The district has total 300.4 thousand ha food grain area which produces total 685.2 thousand ton food grain. The district grows crops like sugar cane, potato etc. besides main cereals and pulse crop. Like in other parts of West Bengal the main cultivators in the district are small and marginal farmers. The availability of quality seeds in this district is a major problem faced by farmers. Most of the poor farmers cannot afford

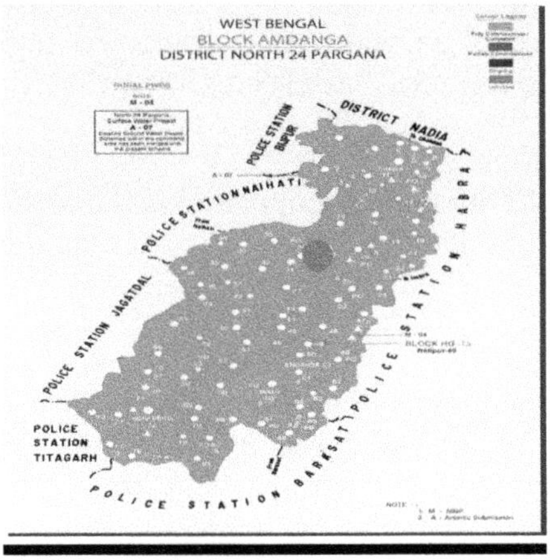

Amdanga block

Population:	191673
Area (Sq.Km):	139.28
Sex ratio (%):	94.3
Cultivators	12754
Agricultural Labourers:	19451
Main Workers:	57433

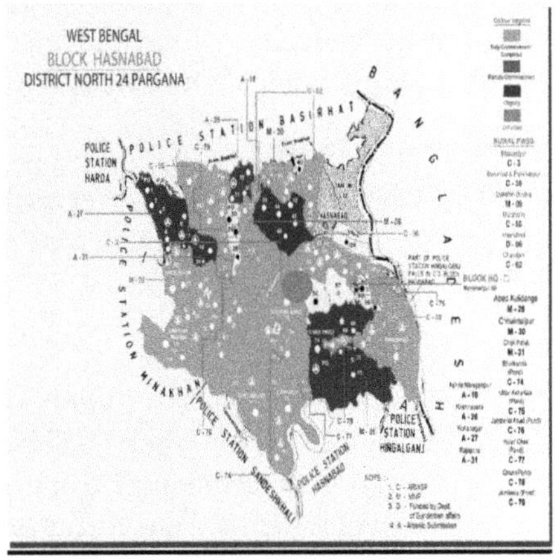

Hansnabad block

Population:	203262
Area (Sq.Km):	153.08
Sex ratio (%):	95.4
Cultivators	8540
Agricultural Labourers:	15729
Main Workers:	53965

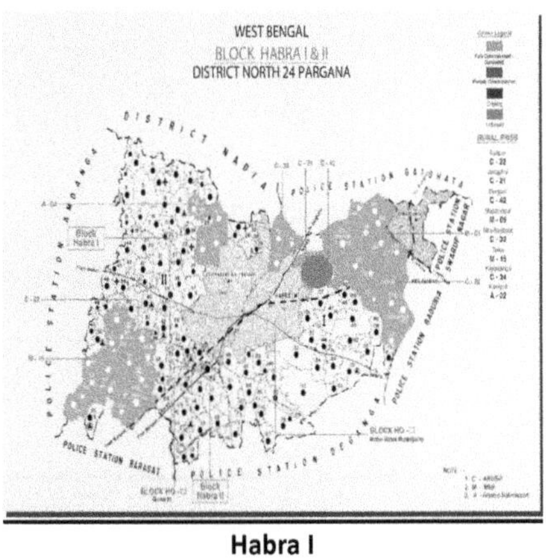

Habra I

Population:	225200
Area (Sq.Km):	117.36
Sex ratio (%)	94.0
Cultivators	7712
Agricultural Labourers	17288
Main Workers:	70535

Habra II

Population:	176490
Area (Sq.Km):	112.68
Sex ratio (%):	95.0
Cultivators	9432
Agricultural Labourers	16527
Main Workers:	54791

Figure 3: CD Block Level Information of North 24 Parganas.

to buy certified seeds at very high prices which come from other states. A sea change has come about in the protection of plants under the Integrated Pest Management Programme. Instead of prescribing use of chemical pesticides,control measures such as cultural,mechanical,varietal,biological and chemical have been given priority.

Land Use Pattern

Cultivable Area of the districts is about 2,71,845 ha. Net cropped Area is 2,64,952 ha. Forest Land is 4,221 ha. Current fallow land is 587 ha. Cutivable wasteland is 465 ha.

Fish Culture

Fishery is an important livelihood and employment generating enterprise in the district of North 24-Parganas. The district has vast inland aquatic resources. The district has good potential for fresh water fish farming. There are about 26007.75 ha of water area of which 8641.45 ha net area is under fish culture. Total 427246 numbers of fishermen are engaged in fish culture activity. Total annual fish production of the district is about 1.82 lakh ton. There are 66 number of Fisheries Co-Operative Society.

Table 1: Demographic and Other Important Features of the North 24-Parganas

Particulars	*Number/percentage*
Gram Panchayat	200
Block	22
Area (sq. km.)	4094.00
Density (per sq. km.)	2445
Total Population	1 0,009,781
Male	5,119,389
Female	4,890,392
Rural	4,277,619
Urban	5,732,162
Total Literary (per cent)	84.1
Male	87.61
Female	80.34

Source: Census of India, 2011

Dakshin Dinajpur

Dakshin Dinajpur is predominantly an agricultural district and represents basically a rural agricultural economy. There are two subdivisions and eight blocks in the district.

West Dinajpur district was created out of erstwhile Dinajpur district in 1947 at the time of partition of India. The rest of the Dinajpur district is now in Bangladesh. The West Dinajpur district was enlarged in 1956. The Dakshin Dinajpur district came into existence on 1st April 1992 through the bifurcation of the erstwhile West Dinajpur district.

Location and Boundary

Dakshin Dinajpur district lies between 26°35′15″ and 25°10′55″ North latitude and 89°0′30″ to 87°48′30″ East longitudes. Its territory extends over an area of 2219 sq. kms. The district is bounded by North Dinajpur district on the West, Malda district on part of the South, while maximum southern and entire eastern and northern boundary is encircled by the International border of Bangladesh (Rajsahi and Dinajpur districts). The district comprises of eight blocks namely Hilli, Balurghat,

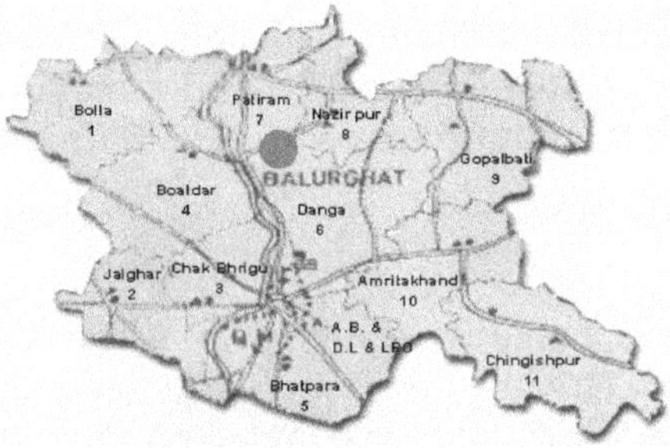

Balurghat Block

Population:	250764
Area (Sq.Km):	368.74
Sex ratio (%):	94.0
Cultivators	26452
Agricultural Labourers	26528
Main Workers:	82559

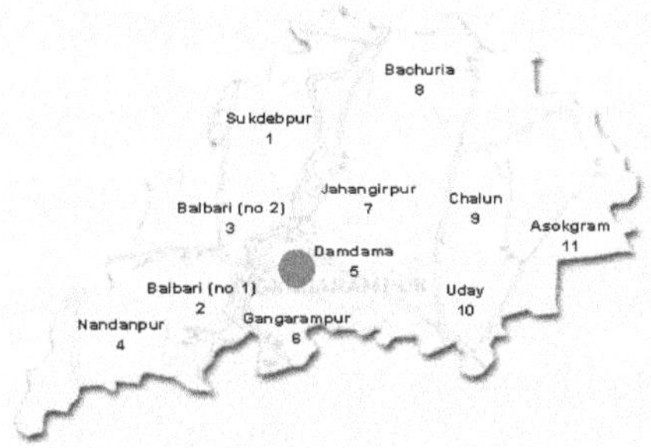

Gangarampur Block

Population:	237628
Area (Sq.Km):	315.52
Sex ratio (%):	93.3
Cultivators	25421
Agricultural Labourers	28929
Main Workers:	76760

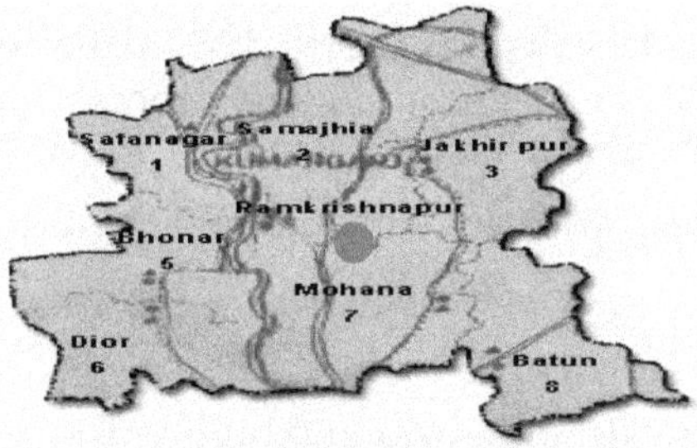

Kumarganj Block

Population:	169102
Area (Sq.Km):	286.63
Sex ratio (%):	94.1
Cultivators	20043
Agricultural Labourers	24573
Main Workers:	54829

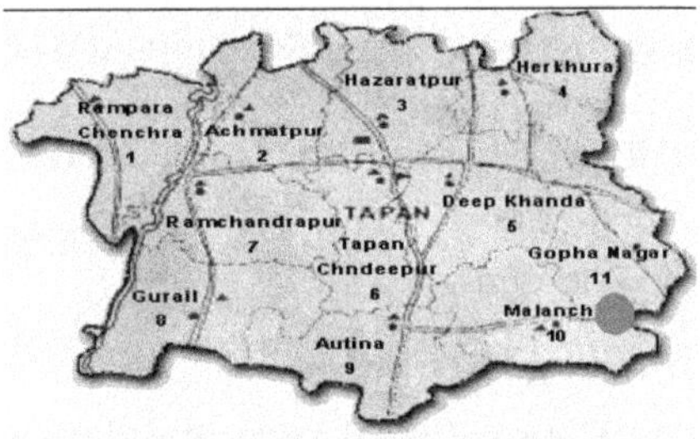

Tapan Block

Population:	250504
Area (Sq.Km):	445.63
Sex ratio (%):	94.9
Cultivators	32687
Agricultural Labourers	31867
Main Workers:	77666

Figure 4: CD Block Level Information of Dakshin Dinajpur.

Tapan, Kumarganj (Balurghat Sub-division), Harirampur, Bansihari, Kushmandi and Gangarampur (Gangarampur Sub-division).

Geology

The soil of the district is mainly of old alluvial type. The soils are light, medium and heavy in texture, upland being lighter and medium to low lands being heavier.

Climate

The district represents the typical climate. It is situated in the sub-tropical climate region. Mainly three seasons that is summer, rainy (monsoon) and winter have been observed. The average annual rainfall is 1690 mm with about 66 rainy days. In the summer season, the maximum temperature is 37°C and minimum temperature is 16°C. In winter season, the minimum temperature is 8°C. The maximum rainfall occurs during July month.

River System

The main rivers of the district are Atrayee, Punarbha, Tangon while Yamuna and Bramhani are small rivers. The main three rivers flow from North to South. The river Atrayee comes from Bangladesh and again goes to Bangladesh through Kumarganj and Balurghat Block. The Punarbha flows through the Gangarampur and Tapan blocks. The Tangon flows through the Kushmandi and Bansihari blocks. During rainy season flood occurs due to over flowing of water. All these rivers are being silted and depth are reducing.

Area and Population

The district of Dakshin Dinajpur was inhabited by 1,676,276 persons of whom 857,199 (51.14 per cent) were men and 819,077 (48.86 per cent) women, the sex ratio worked out to 956 women per 1000 men as per 2011 census. The urban population of the district as per 2011 census was 236,295 (14.09 per cent of the total district population) distributed among the two towns of the district. The rural population of the district was 1,439,981 (85.91 per cent of the total district population), according to the 2011 census, distributed among 1582 inhabited villages.

In terms of area, Dakshin Dinajpur ranks seventeenth among the 19 districts of West Bengal.

Working Population

The total workers of the district was 702,987 and accounted for about 41.94 per cent of the total population of the district. The number of farmer population were about 193,276 (27.49 per cent of total workers). As per 2011 census, the total agricultural labour was 279,932 (which accounted about 39.82 per cent of total workers).

Transport

Transport and communication facilities are not very satisfactory. New rail way line has been laid between Eklakhi and Balurghat, the district headquarter. Train services has been started on 30.12.2004. There is one state highway only 3 km of national highway No. 34 falling within the district.

Agriculture

Agriculture is the major sector of the district. During kharif season rice is the major crop. Other important crops grown in pre-kharif season are aus rice, jute and vegetable and in rabi season potato, mustard, vegetables, wheat, pulses, groundnut and sesame are the main crops.

The Land and Soil Fertility Status

Dakshin Dinajpur district is one of the district having been engulfed with several rivers and so all have considerable bearing to alluvial soils. These soils are named over old alluvium soils as they resemble Bindhya Alluvium soils and also are farthest from river Gangas. Many parts of state fall under the alluvium as for example parts of Burdwan, Hoogly, Midnapore etc. But the largest area coverage seemed to be indulging in Dakshin Dinajpur district excepting the Tapan block having been in lateritic soil group. Generally Old Alluvium of this comprises 70 to 80 per cent of the total area in each blocks of the district excepting Tapan block where 40 per cent alluvium and 60 per cent lateritic of the total area.

Varying textures are characteristics of Dakshin Dinajpur soil. The soils are highly acidic (pH 4.0) to needful (pH 7.0) in some patches. Because of the acidity, the crops are prone to some hazards in different parts of these districts.

Table 2: Demographic and Other Important Features of the Dakshin Dinajpur

Particulars	*Number/percentage*
Gram Panchayat	65
Block	8
Area (sq. km.)	2219.00
Density (per sq. km.)	755
Total Population	1,676,276
Male	857,199
Female	819,077
Rural	1,439,981
Urban	236,295
Total Literary (per cent)	72.82
Male	78.37
Female	67.01

Source: Census of India, 2011.

Topography

Regarding the topographical situation of land forms, mostly the blocks were dominated by medium to low land. In Tapan block the undulating topography reveals the higher high land in comparison to low and medium land. The river bed is generally shrinking and the erosion problem is a predominant problem, as heavy rain causes flash floods in the bank Side. In generally flood prone areas are found in Gangarampur block (heavy higher low land areas) and in Tapan block.

Fish Culture

Dakshin Dinajpur district is enriched with huge aquatic resources. The district has good potential for fresh water fish farming. There are about 8260 hectare of net area for fish culture. The beels, jheels and dighi and other confined water body found in this district mostly resulted from defunct rivers like the Atrayee, Punarbhaba, etc. which used to deposit highly fertile silt in such areas. Total 69953 numbers of fisher folk are engaged in fish culture activity. Total annual fish production of the district is about 5.48 lakh ton.

Chapter 4

Research Methodology

The systematic planning and conduct of a piece of research programme demands an appropriate research methodology, which is a vital prerequisite of any successful research study, since it has in good measure a direct bearing on the relevance and validity of the research findings. In the domain of social science, again it is essential to use a standard method of research design and appropriate techniques of measurement of selected variables and rules or procedure for the testing of observations.

This chapter deals with the various methods and techniques employed in conducting the present research work. The text has been presented under the following heads:

- ✰ Locale of the study
- ✰ Sampling design
- ✰ Variables and their measurement
- ✰ Tools and techniques of data collection
- ✰ Statistical tools

Locale of the Study

The present study was carried out in two districts namely North 24-Parganas and Dakshin Dinajpur district of West Bengal, India. The two districts were selected purposively in view of the vast and diverse inland fishery resources ideally suited for taking up scientific fish culture and they have been situated in different agro ecological zones. Apart from this intrinsic merit of the locale of the study what influenced the researcher in his choice was the additional fact of his familiarity, with the local dialect and accents and intimate knowledge of the social milieu which served as a great help in building up a rapport with the fish farmers, facilitating a quick and easy approach and access to the sources of information. Moreover, judging the condition of its natural and human resources, the district has relatively high potential for conducting research on factors affecting the attitude and adoption behaviour of fish farmers towards scientific fish culture.

Sampling Design

North 24-Parganas district comprises twenty two Development Blocks and Dakshin Dinajpur district comprises eight Development Blocks. Among twenty two Development Blocks in North 24-Parganas four blocks namely Amdanga, Hasnabad, Habra-I and Habra-II and eight Development Blocks in Dakshin Dinajpur district four blocks namely Balurghat, Kumarganj, Tapan and Gangarampur were selected purposively for the study in consideration of the preponderance of fish farmers among the population, coupled with their relatively considerable development potential in the line of economic activity in that though the present level of production and productivity is rather low, there is a good deal of room for the improvement thereof with the introduction of improved, scientific methods and practices of fish culture in place of the present backward ones.

Selection of Villages

From each block, two villages were selected with the help of a random number table following the method of simple random sampling without replacement technique. In total eight villages of each district are served as the representing unit for this study.

Selection of Respondents

The nature of the study demanded that the respondents should be persons engaged in fish culture in addition to their main agricultural profession. Hence, the selected villages were visited by the researcher personally and list of fish farmers were prepared, applying the above criteria.

North 24-Parganas

Villages	*Block*	*Population*		*Sample*
		No. of Fish Farmers	*Proportion in each Village*	*Number of Farmers Selected from each Village*
Chandigarh	Amdanga	132	0.09	120 x 0.09=11
Tababaria		156	0.11	120 x 0.11 =13
Sulkani	Hasnabad	350	0.24	120 x 0.24= 29
Ghosalati		466	0.32	120 x 0.32 =39
Mathurapur	Habra-I	78	0.06	120 x 0.06= 07
Janaphul		95	0.07	120 x 0.07= 08
Suria	Habra-II	86	0.06	120 x 0.06= 07
Guma		72	0.05	120 x 0.05 =06
			Total :	**120**

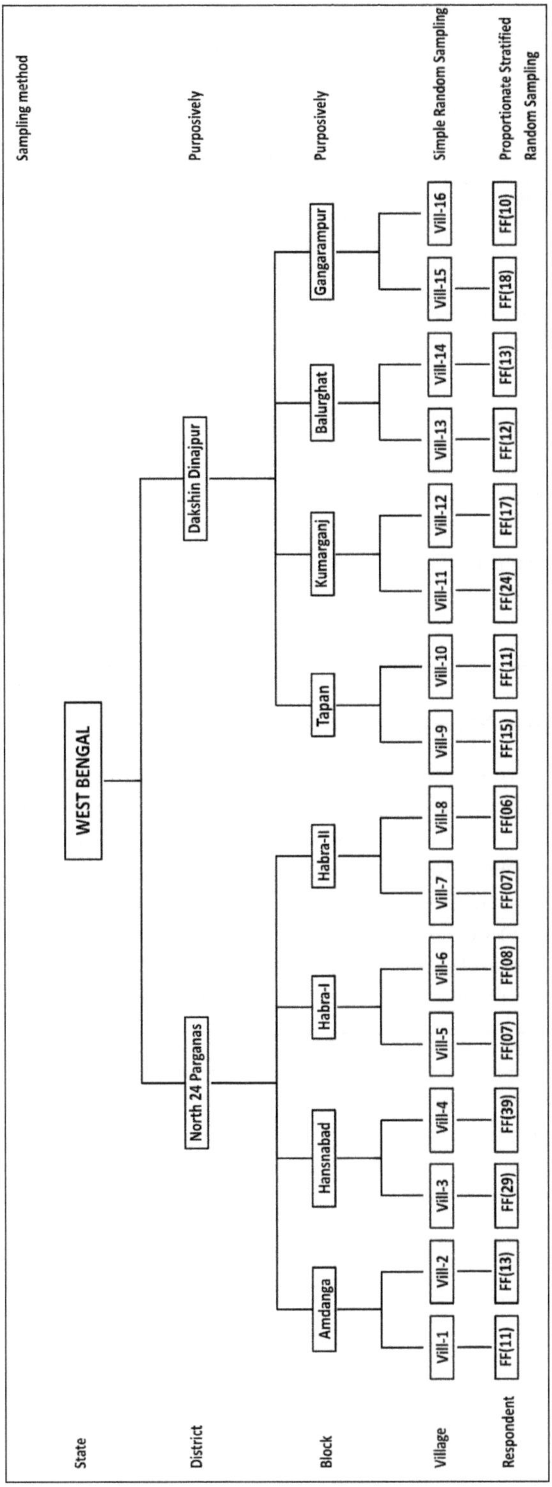

Figure 5: Sampling Plan.

The selected villages of each district which formed the universe were stratified on the basis of the concentration of the fish farmers. A stratified random sample with proportional allocation from each stratum was drawn. Total sample size is 240, of which 120 for each district.

Dakshin Dinajpur

Villages	*Block*	*Population*		*Sample*
		No. of Fish Farmers	*Proportion in each Village*	*Number of Farmers Selected from each Village*
Kashitara	Tapan	34	0.13	120 x 0.13 = 15
Mohukuri		24	0.09	120 x 0.09 = 11
Mohana	Kumarganj	54	0.2	120 x 0.2 = 24
Amritapur		36	0.14	120 x 0.14 = 17
Manipur	Balurghat	26	0.10	120 x 0.10 = 12
Parbatipur		30	0.11	120 x 0.11= 13
Thangapara	Gangarampur	39	0.15	120 x 0.15 = 18
Jalalpur		22	0.08	120 x 0.09 = 10
			Total :	**120**

Variables and their Measurement

Keeping the objectives in view, the adoption behaviour of the fish farmers and their attitude towards scientific fish culture practices were considered as dependent variables. Based on review of literature and discussion with scientists, twenty-four independent variables (personal, socio-economic, psychological, communicational and situational) were included in the present study. The list of variables and their empirical measurements used in the study are presented below. The scales developed by earlier researchers for measurement of variables have suitably been modified and used in the present study wherever required.

Sl.No.	*Variables*	*Measurements*
I.	**Dependent variables**	
1.	Attitude of fish farmers towards scientific aquaculture	Scale developed by Nagarajaiah (2002)
2.	Adoption behaviour of fish farmers towards scientific fish culture	Scoring procedure developed by Sinha and Kolte (1974)
II.	**Independent variables**	
A.	***Socio-personal variables***	
1.	Age	Chronological age in years
2.	Caste	Structured schedule developed for the study
3.	Family size	Structured schedule developed for the study

Contd...

Contd...

Sl.No.	Variables	Measurements
4.	Family type	Structured schedule developed for the study
5.	Education	Structured schedule developed for the study
6.	Fish farming experience	Structured schedule developed for the study
B.	***Socio-economic variables***	
1.	Occupation	Structured schedule developed for the study
2.	Annual income	Structured schedule developed for the study
3.	Land holding	Structured schedule developed for the study
4.	Social participation	Structured schedule developed for the study
5.	Possession of fishing equipment	Structured schedule developed for the study
C.	***Communicational variables***	
1.	Mass media participation	Procedure followed by Patil (1990)
2.	Extension agency contact	Structured schedule developed for the study
3.	Cosmopoliteness	Scale developed by Moulik (1965)
D.	***Psychological variables***	
1.	Innovative proneness	Scale developed by Moulik and Rao (1973)
2.	Credit orientation	Procedure followed by Beal and Sibley (1967)
3.	Value orientation	Scale developed by Murthy and Singh (1969)
4.	Risk orientation	Scale developed by Supe (1969)
5.	Economic motivation	Scale developed by Supe (1969)
6.	Knowledge about scientific fish culture	Teacher made test
E.	***Situational variables***	
1.	Pond size (area)	Structured schedule developed for the study
2.	Duration of water availability	Structured schedule developed for the study
3.	Source of water	Structured schedule developed for the study
4.	Extent of Weed Infestation	Structured schedule developed for the study
III.	**Descriptive Variable**	
	Constraints to adoption of scientific fish culture	Structured schedule developed for the study

Operationalization and Measurement of Dependent Variables

Attitude towards Scientific Culture Practices

Thurstone (1946) defined attitude as the degree of positive or negative affect associated with some psychological object. It may be any symbol, phrase, slogan, individual, institution, idea towards which people can differ with respect to positive or negative effect.

In the present study, attitude has been operationalized as respondents' degree of favourableness or unfavourablness towards different aspects of scientific fish culture practices. Scientific fish culture practices aim at efficient use of water body by adopting scientific fish culture practices to increase the income to fish farmer by increased fish production and this also creates additional rural employment for the unemployed youths by resorting to fish farming activity in village tanks. One of the baffling problems that policy makers face is fish farmers' attitude towards various components of developmental programme which are assured to increase fish production. In the present study, scale developed by Nagarajaiah (2002) have been utilized to measure attitude towards scientific fish culture practices which help to educate the clientele in the right direction. The method suggested by Likert (1932) in developing summated ratings has been followed in the present study.

Construction and Standardization of Attitude Scale

Item collection : The first step in the construction was to collect a large number of statements covering different dimensions of composite fish culture practices. The statements were prepared after a review of literature and discussions with specialists, scientists, fisheries departmental officers, extension personnel, fish farmers and bankers. Further, extant literature on composite fish culture technology as well as on the objectives and working of Fish Farmers Development Agencies for promotion of composite fish culture technology was collected on an extensive scale from different sources, including various published literature and Fisheries Department, Govt. of India. By this process eighty four statements were prepared. The statements thus prepared were carefully edited in the light of the 14 criteria suggested by Edwards (1969). After such screening only fifty one statements out of eighty four were finally retained. In order to arrive at a comprehensive scale to measure the attitude towards different aspects of composite fish culture practices, the selected statements were grouped under four broad components considering the different dimensions. The components identified were

i) Objectives and feasibility
ii) Technology transfer and development
iii) Peoples participation
iv) Production and risk factors

In all fifty one statements were retained in the draft scale.

Relevancy of Statements

All the selected statements (hereinafter) referred to as items were then sent to the 120 experts/judges *viz.* extension specialists, Professors/Associate Professor of State Agricultural Universities/ICAR Institutes and persons with considerable practical experience in the promotion of composite fish culture technology for the purpose of determining their relevancy and subsequent screening of items for their inclusion in the final scale. The judges were requested with appropriate instructions critically to evaluate the relevancy of each item in measuring the attitude of fish farmers and to give their response on a four point continuum *viz.* more relevant, relevant, less relevant and not relevant with the corresponding scores of 4, 3, 2 and 1 respectively. The judges were also requested to make necessary modification and addition or deletion of items if they desired so.

The responses were obtained from 51 judges in time. Out of 51, the responses of 3 judges were excluded as being incomplete. Finally, responses from 48 judges were considered for further analysis.

Final Selection of Statements

Each statement was weighed on the basis of relevancy weightage for deciding its inclusion in the scale. *'Relevancy weightage'* is defined as the ratio of actual score obtained by a statement to the maximum possible score that the statement could secure. By definition the relevancy weightage can range from 0.25 to 1.00. The criterion laid down for the selection of items was 0.65 and more than 0.65 relevancy weightage. By this process 38 items were isolated and retained in the first stage. Further, in the light of criticisms, suggestions and comments of judges, these selected items were modified and rewritten. The number of items, qualified under each component were 10 out of 13 for objectives and feasibility. 10 out of 13 for technology transfer and development, 9 out of 13 for people's participation and 9 out of 12 for production and risk factors.

Preliminary administration of the scale : The items were translated into local dialect and were administered personally by interviewing 30 fish farmers from the non-sample area. Responses for the items were obtained on five point continuum *viz.* 'strongly agree', 'agree', 'undecided', 'disagree' and 'strongly disagree' with the scores of 5, 4, 3, 2 and 1, respectively for favourable items and reverse scoring for unfavourable items. Attitude score of the respondent was obtained by summing up the scores of all the items.

Item Analysis

The purpose of item analysis was to examine how well each item discriminates the person with highly favourable attitude from the person with least favourable attitude in question. For this purpose, the respondents attitude scores were arranged in descending order. Twenty five per cent of the respondents with highest total scores and 25 per cent of the respondents with lowest total scores were selected. These two groups served as the criterion groups in terms of which the item analysis was conducted. The critical ratio (r) was calculated by the following formula.

$$t = \frac{\overline{X}_H - \overline{X}_L}{\sqrt{\dfrac{\{(\sum X^2_H - \dfrac{(\sum X_H)^2}{n}\} + \{(\sum X^2_L - \dfrac{(\sum X_L)^2}{n}\}}{n(n-1)}}}$$

where,

$\overline{X}_H$ = The mean score on a given item for high score group

$\overline{X}_L$ = The mean score on a given item for low score group

$\sum X^2_H$ = Sum of the squares of the individual scores in the high score group

ΣX^2_L = Sum of the squares of the individual scores in the low score group

n = Number of respondents in each group

The 't' value is a measure of the extent to which a given item differentiates the high group from the low group. Appendix-III contains the list of 38 items along with calculated 't' values. As a crude and approximate rule of thumb, it was decided to select all those items which obtained 't' value of 1.75 and above (> 1.75) as suggested by Edward (1969). Based on this criterion, 20 statements which qualified for inclusion in the final scale were selected.

Thus, the final attitude scale consisted of 20 items; 4 under objectives and feasibility, 5 under technology transfer and development, 6 under people's participation and 5 under production and risk factors. These statements covered all the aspects of the attitude of the fish farmers under study towards composite fish culture practices.

Attitude Aspect	*No. of Statements Retained after Initial Screening*	*No. of Statements Qualified in Relevancy Test (Rw)*	*No. of Statements Qualified in Critical Ratio Test (t)*
Objective and feasibility	13	10	4
Technology transfer and development	13	10	5
people's participation	13	9	6
Production and risk factors	12	9	5
Total :	**51**	**38**	**20**

Validity of Scale

Validity refers to whether the instrument measures the same variable/aspects which it is supposed to measure. In the present study, content validity was justified by seeking the opinion of the experts in the field concerned. Further, due care was exercised in obtaining a fair degree of content validity by covering all the relevant dimensions. Further, based on the response obtained from 30 respondents in the non-sample area, item analysis was carried out and 't' value for each statement was calculated. Based on the t' value, only qualified statements were selected and included in the final scale ensuring a fair degree of content validity.

Reliability of the Scale

A scale is said to be reliable when it produces results with a high degree of consistency when administered to the same respondents at different point of time. In the present study 'split half' method (Henry *et al.*, 1973) was used to establish the reliability of the scale. The scale was split into two halves on the basis of odd and even number of items and administered to 30 respondents of non-sample area. Thus, two sets of scores were obtained. A Pearson's product moment correlation coefficient was calculated between two sets of scores obtained, using the following formula.

Coefficient of internal consistency (rα) =

$$\frac{N\sum XY-(\sum X)(\sum Y)}{\sqrt{[N\sum X^2-(\sum X)^2][N\sum Y^2-(\sum Y)^2]}}$$

where,

N = Number of respondents

X = Value of odd numbered items scores

Y = Value of even numbered items scores

The value rα obtained was 0.74, and this was again got corrected by using the Spearman Brown formula, and thus obtained the reliability of the original test. The formula used was

$$r\alpha=\frac{2r\alpha}{1+r\alpha}$$

The rα obtained value was 0.8505 which indicated a high reliability of the scale.

Method of Scoring

The final scale consisting of 20 items was administered to the respondents in the study area. The response was obtained on a five point continuum *viz.*, 'strongly agree' (SA), 'agree' (A), 'undecided' (UD), 'disagree' (DA) and 'strongly disagree' (SDA) with weightage of 5, 4, 3, 2 and 1 for positive items while reverse scoring system was employed for negative items. The total attitude score for each respondent was calculated by adding the score of all the statements. The possible total score of the scale ranges from 100 to 20. Based on the scores obtained by the respondents, they were categorized into three categories, taking mean and standard deviation as measures of check.

Category	*Scores*	
	North 24 Parganas	*Dakshin Dinajpur*
Least favourable (up to X- ½ SD)	Up to 63.25	Up to 58.83
Favourable (X- ½ SD to X+ ½ SD)	63.25 to 69.0	58.83 to 67.36
More favourable (above X+ ½ SD)	Above 69.0	Above 67.36
	Mean : 66.13 SD : 5.75	Mean : 66.13 SD : 5.75

Adoption Behaviour of Fish Farmers towards Scientific Fish Culture Practices

According to Rogers (1962) 'adoption' is a decision to continue full use of an innovation. Das Gupta (1989) defined adoption as the integration of an innovation into a farmers' normal farming activity over an extended period of time. Further, he observed that at the stage of adoption, the farmer integrates the innovation into his system of farming operation and it involves continued full scale use of an innovation by the farmer until he decides to replace it with a better practice or discontinue its use for some other reasons. Thus, adoption can be termed as behavioural response. It is the overt behaviour of a farmer expressed in terms of aggregate adoption scores obtained by him with respect to recommended practices.

In the present study, the extent of adoption was conceptualized as the extent to which the fish farmers (respondents) have adopted the recommended composite fish culture practices for increased fish production.

The methodology followed in developing scoring procedure for measuring adoption of composite fish culture practices :

In the present study, the extent of adoption of recommended composite fish culture practices was measured by using the adoption quotient developed by Sen Gupta (1967). Accordingly, the following formula was used to calculate the general adoption level.

$$\text{Adoption quotient} = \frac{\text{Adoption score of the respondent}}{\text{Maximum adoption score one could get}} \times 100$$

The partial adoption technique suggested by Sinha and Kolte (1974) was followed with necessary modification for scoring the practices partially adopted by the respondents. The following scores were assigned respectively for full, partial and non-adoption of the recommended practices.

Adoption Level	*Score*
Full adoption	2
Partial adoption	1
Non-adoption	0

The fact of partial adoption was determined by taking into cognizance of any deviation from the recommendation. The raw adoption scores were converted into adoption quotient for measuring the general adoption level.

Thus, after computing individual adoption quotient scores, the respondents were grouped into three categories with mean and standard deviation as measure of check.

Category	*Scores*	
	North 24 Parganas	*Dakshin Dinajpur*
Low adoption (up to X- ½ SD)	Upto 72.34	Up to 61.25
Medium adoption(X- ½ SD to X+ ½ SD)	72.34 to 80.81	61.25 to 73.77
High adoption (above X+ ½ SD)	Above 80.81	Above 73.77
	Mean : 76.57 SD : 8.49	Mean: 67.51 SD : 12.52

Selection of Recommended Composite Fish Culture Practices

Taking into consideration the fundamental principles of composite fish culture technology, improved fish culture practices recommended for adoption in irrigation tanks by operating FFDAs and department of fisheries in West Bengal were identified. These identified practices were considered after an in-depth detailed review of literature published by FFDAs. Directorate of Fisheries, Central Inland Fisheries Research Institutes, University of Agricultural Sciences and also discussion with the officials and scientists in the concerned field who are working with the fish farmers. By this process 17 improved fish culture practices were selected as relevant to the study area. The practicability of the recommended fish culture practices was got further confirmed by discussions with the concerned scientists, the extension personnel of the FFDAs and officials of Department of Fisheries. The criteria like 'recommendation of practices for longer period (many years)', 'fair degree of communication and extension efforts by the implementing agency' and 'general acceptance of the practice among fish farmers' were also kept in view while finalizing the identification of the practices. The 17 practices which were finally selected for the study are mentioned below along with the respective scores assigned for each practice.

Recommended Practices	*Scores Assigned for*		
	Full Adoption (2)	*Partial Adoption (1)*	*Non-adoption (0)*
A. Pre-stocking practices			
1. Testing of soil and water Recommendation : Testing for water and soil for acidity/alkalinity	2	1	0
2. Control of aquatic weeds Recommendation : Removal/eradication of aquatic weeds by chemical/manual method	2	1	0
3. Control of predators and undesirable fish fauna Recommendation : Complete eradication using piscicides in recommended levels or repeated netting or any other method	2	1	0
4. Application of lime for improving soil and water quality F.A. : Recommended dosage (250-300 kg/ha) PA : Less than recommended dosage NA : No application of lime	2	1	0
5. Initial manuring with organic manures (RCD or poultry) FA : Recommended dosage (2500 kg/ha) PA : Less than recommended dosage	2	1	0
6. Application of inorganic fertilizers (urea and super phosphate) FA : Recommended dosage (25 kg urea + 50 kg SPO_4 kg/ha) PA : Less than recommended dosage	2	1	0
7. Period of application of initial manures and fertilizers before stocking FA : 7 to 15 days before stocking PA : 3-6 to 16-20 days before stocking NA : Any other period	2	1	0
B. Stocking practices			
8. Stocking of recommended number of different fish species FA : 4 to 6 species PA : 1 to 3 species	2	1	0

Contd...

Contd...

Recommended Practices	Scores Assigned for		
	Full Adoption (2)	Partial Adoption (1)	Non-adoption (0)
9. Density of stocking of fish seed (no/ha) FA : 4001 to 5000 no/ha PA : < 4000 no/ha	2	1	0
10. Ratio of stocking of selected species FA : As recommended ratio PA : Other than recommended ratio NA : Single species stocking	2	1	0
11. Size of the fish seed stocked FA : More than 30 mm PA : 20-30 mm NA : Less than 20 mm	2	1	0
C. Post-stocking practices			
12. Use of supplementary feeds in prescribed proportion FA : Recommended proportion of GOC and RB PA : Use of single ingredient or deviations in the prescribed ratio of GOC and RB or inclusion of any local feeds etc. in the feed NA : No feeding	2	1	0
13. Rate of feeding FA : Recommended rate (2 per cent of body weight/day) PA : Less than recommended rate	2	1	0
14. Feeding method FA : Recommended method (bag/tray feeding/lump method) PA : Other than recommended method	2	1	0
15. Post-stocking manuring with organic manures FA : Recommended dosage (250 kg/ha/month) PA : Less than recommended dosage	2	1	0
16. Post-stocking fertilization with inorganic fertilizers FA : Recommended (30-40 kg/ha/month) PA : Less than recommended	2	1	0
17. Periodic netting to check the growth and health of fishes FA : Recommended (monthly netting) PA : Other than recommended (netting with varying frequency) NA : No netting	2	1	0

The overall adoption score of a respondent is the sum total of scores obtained for each practice mentioned above. The maximum and minimum overall adoption score that a respondent might secure was 34 and 0, respectively.

Operationalization and Measurement of Independent Variables

Age

Age was operationalized as the number of years completed by the respondent at the time of investigation. The fish farmers were categorized into three groups according to their age.

Category	Score
Young	upto 30 years
Middle	31 to 45 years
Old	46 years and above

Caste

The caste of the respondents was recorded and classified as follows. The procedure followed by Shashikumar (1998) was adopted for classification and scoring procedure adopted is as follows:

Category	Score
Scheduled Tribes	1
Scheduled Castes	2
General	3

Family Size

It refers to a group of closely related persons living together under a single roof and was measured by the procedure followed by Prasad (1995). The respondents were classified into small and large families. Scores of one and two were given respectively for quantifying the categories small and large.

Category	No. of Family Members
Small	Up to 5 members
Large	More than 5 members

Family Type

It refers to the fact that whether the respondents belongs to the nuclear (single family) or joint family. Nuclear or single family refers to a couple and their unmarried children living together. Joint family refers to any additional member who is economically independent living with the nuclear family. Based on the family type, the respondents were categorized into two classes. Scores of one and two were assigned respectively for the nuclear and the joint family categories.

Category	Score
Nuclear	1
Joint	2

Education

It refers to the number of years spent on formal education acquired by the respondent. The scoring patterns for the various categories of educational level as followed in the present study is given below.

Education Level	Score
Illiterate	0
Can read and write	1
Primary school	2
Middle school	3
Secondary	4
Higher Secondary	5
Graduate	6

Frequency and percentage were taken into consideration while classifying the respondents into different categories.

Fish Farming Experience

It refers to the number of years already spent by the respondent in fish farming activities at the time of enquiry. Each year of experience in fish farming was assigned the score of one. On the basis of scores obtained the respondents were categorized into three categories, according to their experience.

Category	Experience
Low	Up to 2 years
Medium	3 to 6 years
High	7 years and above

Socio-economic Variables

Occupational Status

It refers to occupation held by the respondent as main occupation or secondary occupation. The procedure developed by Meeran (1983) was used in the present study with a slight modification. The scoring procedure adopted is as follows :

Category	Score
Fish culture as main occupation	2
Fish culture as secondary occupation	1

Annual Income

It refers to the total income earned by the respondent from all his gainful occupations, including agriculture, fish farming, fishing and any other economic enterprises as practised. The respondent's total annual income as rounded off in one lakh rupees was used as annual income score. It was categorized according to the classification suggested by Ministry of Rural Development, Govt. of India (Anon, 1992) as indicated below :

i) Up to Rs. 0.11 lakh

ii) Rs. 0.11 lakh to Rs 0.22 lakh
iii) Rs. 0.22 lakh to Rs 0.33 lakh
iv) Above Rs 0.33 lakh

Size of Land Holding

This refers to the extent of land possessed by the respondent for cultivation. Data pertaining to the total number of acres of dry land, irrigated and garden land as reported by the respondents as being under his possession were collected. The extent of the wet and garden lands were converted into dry land units by multiplying the area of the irrigated land, wet or garden by two. Then this acreage was added with dry and thus all types of lands were expressed in the form of dry land. The total area of dry land in acres was the size of land holding possessed by the respondent. The respondents who possess more than 5 acres were categorized as big farmer, those who possess up to 5 acres as small farmers and those who possessed less than 2 acres of land as marginal farmer.

Sl.No.	Category	Extent of Dry Land Owned in Standard Acres
1.	Marginal farmers	Up to 2 acres
2.	Small farmers	2 to 5 acres
3.	Big farmers	More than 5 acres

Social Participation

It was operationalized as the degree of involvement of the respondent in any formal and/or informal social organization as a member or office bearer in his/her village or nearby places. Social participation was empirically measured by using the procedure outlined by Patil (1990) with a slight modification.

Organization	Non member	Member	Office Bearer	Frequency Participation in Activities		
				Regularly	Occasio-nally	Never
Gram Panchayat/Zilla Parishad	0	1	2	2	1	0
Youth club/Mahila Mandal	0	1	2	2	1	0
Farmers' forum	0	1	2	2	1	0
Co-operative (including FCS)	0	1	2	2	1	0
Others	0	1	2	2	1	0

The composite score was arrived at by summing up the scores obtained by the respondent on each item listed. Higher score reveals greater involvement of individual in organization as a member/office bearer. Further the respondents were classified into three categories based on mean and S.D. as a measure of check.

Category	Scores	
	North 24 Parganas	Dakshin Dinajpur
Low social participation (up to X- ½ SD)	Up to 3.5	Up to 3.19
Medium social participation (X- ½ SD to X- ½ SD)	3.5 to 6.4	3.19 to 5.7
High social participation (above X+ ½ SD)	Above 6.4	Above 5.7
	Mean : 4.94 SD : 2.89	Mean : 4.47 SD : 2.57

Possession of Fishing Equipments

In the present study, it refers to possession of the different kinds and types of gears/fishing equipments like fishing nets etc. and used in fish culture and harvesting. Scores of zero and one were assigned respectively for no net and having net categories.

Category	Score
No net	0
Caste nets/drag net	1

Communicational Variables

Mass media participation : It refers to the participation of individuals to different mass communication media in the related activities such as listening to radio, viewing television, reading newspapers, farm magazines and leaflets, etc. The quantification of this variable was done according to the procedure followed by Patil (1990) with a slight modification.

The maximum and the minimum score for each respondent were 16 and zero respectively. Higher scores reveal greater participation in mass media by the respondent. Further, respondents were categorized into three categories based on mean and standard deviation as measure of check.

Category	Scores	
	North 24 Parganas	Dakshin Dinajpur
Low (up to X- ½ SD)	Up to 5.6	Up to 4.7
Medium (X- ½ SD to X+ ½ SD)	5.6 to 8	4.7 to 7.6
High (above X+ ½ SD)	Above 8	Above 7.6
	Mean : 6.86 SD : 1.22	Mean : 6.16 SD : 2.94

Extension Agency Contact

In the present study it precisely refers to the weekly frequency of the contact of the fish farmers with various extension agencies like FEO, Fishery Assistant,

Scientists, Progressive farmers, for the purpose of collecting information regarding fish culture practices.

The responses were obtained on 4 point continuum: frequently, often, seldom and never and scores were assigned as follows –

Category	*Weekly Frequency*	*Score*
Frequently	6 times	3
Often	1-6 times	2
Seldom	1-3 times	1
Never	0	0

Thus, the sum total of the scores obtained by a fish farmer in respect of all the extension agencies with whom he/she is reportedly in contact was considered as the extension agency contact score for him. Further, respondents were categorized into three categories based on mean and standard deviation as measure of check.

Category	*Scores*	
	North 24 Parganas	*Dakshin Dinajpur*
Low (up to X- ½ SD)	Up to 4.02	Up to 2.97
Medium (X- ½ SD to X+ ½ SD)	4.02 to 6.02	2.97 to 5.29
High (above X+ ½ SD)	Above 6.02	Above 5.29
	Mean : 5.02 SD : 2.0	Mean : 6.16 SD : 2.94

Cosmopoliteness

A cosmopolite individual has some interest in the community and maintains a minimum relationship within the community, is also highly oriented towards problems and affairs outside the community and regards himself as a part of the outside world.

To assess the cosmopoliteness, of a respondent the scale developed by Moulik (1965) was used. The respondents were asked to rate each statement on a three point continuum, namely, agree (A), undecided (UD) and disagree (DA). For positive statements *i.e.* (2 and 4) scores of 2, 1 and 0 were assigned and a reverse scoring system was followed for negative statements *i.e.* (1 and 3) respectively. Further, the respondents were categorized into three categories based on mean and standard deviation as measure of check.

Category	*Scores*	
	North 24 Parganas	*Dakshin Dinajpur*
Low (up to X- ½ SD)	Up to 5.39	Up to 4.53
Medium (X- ½ SD to X+ ½ SD)	5.39 to 7.11	4.53 to 6.8
High (above X+ ½ SD)	Above 7.11	Above 6.8
	Mean : 6.25 SD : 1.27	Mean : 5.67 SD : 2.27

Psychological Variables

Innovative Proneness

It was operationalized as the degree to which a farmer is eager in adopting the innovations early in his fish culture pond.

In this study, the fish farmers' innovative proneness was measured by using the scale constructed by Moulik and Rao (1973) with a little modification to suit the present investigation. The scale consists of five statements, out of which statements number 1 and 4 were positive statements, while statements number 2, 3 and 5 were negative statements. The responses for each statement were noted on a five point continuum, namely, SA, A, UD, DA and SDA with the score of 5, 4, 3, 2 and 1 for positive statements and 1, 2, 3, 4 and 5 for negative statements respectively. The maximum score an individual could obtain on this scale was 25 and the minimum was 5. Further, respondents were categorized into three categories, taking mean and standard deviation as measure of check.

Category	*Scores*	
	North 24 Parganas	*Dakshin Dinajpur*
Low (up to X- ½ SD)	Up to 16.16	Up to 14.19
Medium (X- ½ SD to X+ ½ SD)	16.16 to 18.71	14.19 to 17.91
High (above X+ ½ SD)	Above 18.71	Above 17.91
	Mean : 17.44 SD : 2.55	Mean : 15.67 SD : 2.27

Credit Orientation

This was operationalized as the favourable and positive attitude of an individual respondent towards obtaining credit from institutional sources for fish farming purposes. The measurement and scoring procedure of Beal and Sibley (1967) with a slight modification was employed. The data is presented below.

Sl.No.	*Items*	*Response*	*Score*
1.	Do you think fish farmer should borrow money for fish culture?	Yes	1
		No	0
2.	How difficult is it, in your opinion, to secure credit for fish culture purposes?	Very easy	3
		Easy	2
		Difficult	1
		Very difficult	0
3.	How is a fish farmer treated when he goes to secure credit?	Very fairly	3
		Fairly	2
		Badly	1
		Very badly	0
4.	There is nothing wrong in taking credit from institutional sources for increasing fish production	Strongly agree	3
		Agree	2
		Disagree	1
		Strongly disagree	0
5.	Did you use credit in the last two years for fish farming?	Yes	1
		No	0

The grand total was arrived at by adding the scores assigned to the respondent in each of the above item. The scores added for all the items formed the credit orientation score of the respondent. The maximum and minimum scores that could be obtained by the respondent were 11 and 0 respectively. The higher the score the more favourable is the attitude of the respondent concerned. Further, the respondents were classified into three groups, taking mean and S.D. as measure of check.

Category	*Scores*	
	North 24 Parganas	*Dakshin Dinajpur*
Low (up to X- ½ SD)	Up to 4.06	Up to 3.78
Medium (X- ½ SD to X+ ½ SD)	4.06 to 5.82	3.78 to 6.09
High (above X+ ½ SD)	Above 5.82	Above 6.09
	Mean : 4.96 SD : 1.76	Mean : 4.94 SD : 2.31

Value Orientation

Adaptation of practices of fish farmers is an expression of transactional act and therefore, depends to some extent on his value orientation. The value orientation of fish farmers was studied in respect of two dimensions: Localite-cosmopolite (Lo-Co) and external conformity – individualism (E-I). The value orientation of fish farmers was measured in this study with the help of the scale developed by Singh and Murty (1969).

There were three negative items in the localite – cosmopolite scale. A strong agreement with the three negative items would indicate localite value orientation while a strong agreement with two positive items would indicate cosmopolite value orientation. In the case of external conformity, individualism scale, a strong agreement with the three negative items would indicate external-conformity and a strong agreement with two positive items would indicate individualism. The scoring techniques is illustrated below –

For Negative Items

Response :	*Strongly Agree*	*Agree*	*Disagree*	*Strongly Disagree*
	(1)	(2)	(3)	(4)

For Positive Items

Response :	*Strongly Agree*	*Agree*	*Disagree*	*Strongly Disagree*
	(4)	(3)	(2)	(1)

Further, the respondents were categorized into three categories taking mean and standard deviation as measure of check.

Category	*Scores*	
	North 24 Parganas	*Dakshin Dinajpur*
Low (up to X- ½ SD)	Up to 24.02	Up to 22.36
Medium (X- ½ SD to X+ ½ SD)	24.02 to 27.13	22.36 to 26.55
High (above X+ ½ SD)	Above 27.13	Above 26.55
	Mean: 25.58 SD : 3.11	Mean : 24.46 SD : 4.19

Risk Orientation

Risk orientation was opetationalized as the degree to which a respondent is oriented towards risk and uncertainties involved in scientific fish culture and has courage to face the problems while practising scientific fish culture.

Risk orientation was measured with the help of the risk orientation scale developed by Supe (1969) with slight modification. The scale consisted of 6 statements. The items were rated in five point response categories ranging from 'strongly agree' to 'strongly disagree'. There were four positive and two negative items. The scoring procedure for positive items were 'strongly agree'-7, 'agree'-5, 'undecided'-4, 'disagree'-3 and 'strongly disagree'-1. The scoring procedure was reverse for the negative responses. By summing up the score recorded; total score was computed. The range of score was from 6 to 42. Further, the respondents were categorized into three categories taking mean and standard deviation as measure of check.

Category	*Scores*	
	North 24 Parganas	*Dakshin Dinajpur*
Low (up to X- ½ SD)	Up to 19.17	Up to 18.2
Medium (X- ½ SD to X+ ½ SD)	19.17 to 21.20	18.2 to 20.82
High (above X+ ½ SD)	Above 21.20	Above 20.82
	Mean : 20.19 SD : 2.03	Mean : 19.51 SD : 2.62

Economic Motivation

Economic motivation was conceptualized as one's orientation towards profit maximization in fish farming activity.

It was measured with the help of the economic motivation scale developed by Supe (1969) with a little modification to suit the needs of the present study. The scale consisted of 6 statements. The items were rated in five point response categories ranging from 'strongly agree' to 'strongly disagree'. There were four positive and two negative items. The scoring procedure for positive items were 'strongly agree'-7, 'agree'-5, 'undecided'-4, 'disagree'-3 and 'strongly disagree'-1. The scoring procedure was reverse for the negative responses. By summing up the score recorded; total score was computed. The range of score was from 6 to 42.

Further, the respondents were categorized into 3 categories, taking mean and SD as measure of check.

Category	Scores	
	North 24 Parganas	Dakshin Dinajpur
Low (up to X- ½ SD)	Up to 20.57	Up to 18.73
Medium (X- ½ SD to X+ ½ SD)	20.57 to 23.37	18.73 to 22.57
High (above X+ ½ SD)	Above 23.37	Above 22.57
	Mean : 21.97 SD : 2.8	Mean : 20.65 SD : 3.84

Extent of Knowledge about Scientific Fish Culture Practices

In this study, Knowledge refers to the body of information understood and retained by the respondents about composite fish culture which comprises various improved fish culture and management practices for increased fish production from a unit area of water body.

Construction of Knowledge Test

For the purpose of the present investigation, a teacher made knowledge test was devised to measure the knowledge of the respondents about the composite fish culture practices. Wadt and Brown (1963) have recommended that teacher made test can be constructed and used for the assessment of knowledge given in outside formal learning situations. Ebel and Damrin (1960) have remarked that the 'short answer tests' are used mainly because of their virtue to test the examinee's ability to produce or recall the answer. Ehatia (1965) opines that teacher made test may meet special needs of the particular content of the subject and of local condition. Further, they are easy to score and may be more effective for a specialized purpose.

The procedure followed in the construction of knowledge test is as follows:

Item Collection

The content of the knowledge test was composed of questions called items. An item pool or questions was prepared by referring to different sources such as literature, extension publications of the Department of Fisheries, discussions with Scientists and Specialists in the field of inland fisheries development and officers of the fisheries department who are involved in extension activities. This process was further supplemented with the review of literature. On the basis of all this, a set of knowledge questions with appropriate answers was prepared in consultation with scientists and experienced departmental officers involved in extension and training programme for fish farmers in the State. The set of knowledge questions so prepared was given to a panel of experts comprising of subject matter specialists, scientists and senior departmental officers who had vast experience in the field for their comments and suggestions. Based on the critical observations and suggestion received from the experts, the knowledge items were modified and new items were added as and where pointed out. Adequate care was taken while finalizing the knowledge questions about their clarity and the items should promote thinking rather than memorization. Finally, 41 knowledge items were retained out, of the 55 items initially listed with possible answers.

Method of Scoring

The knowledge test consisted of 41 items covering different aspects of composite fish culture practices, *i.e.*, pre-stocking, stocking and post-stocking practices with required management skills. All the items were objective type and the 'one score was given for the 'yes' response and 'zero' for 'no' response. The summation of scores for the 'yes' responses over all the items for a particular respondent indicated his extent of knowledge. The maximum and the minimum score would be forty one and zero respectively. Further, the raw knowledge score of each individual respondent was converted into a knowledge index, using the formula.

$$\text{Knowledge index (KI)} = \frac{\text{Number of correct responses}}{\text{Total number of knowledge items}} \times 100$$

Thus, after computing knowledge index scores, the respondents were categorized into three categories taking mean and standard deviation as a measure of check.

Category	*Scores*	
	North 24 Parganas	*Dakshin Dinajpur*
Low (up to X- ½ SD)	Up to 72.35	Up to 65.65
Medium (X- ½ SD to X+ ½ SD)	72.35 to 81.27	65.65 to 79.63
High (above X+ ½ SD)	Above 81.27	Above 79.63
	Mean : 76.81 SD : 8.92	Mean : 72.64 SD : 13.98

Situational Variables

Size of Water Body

It refers to the extent of water spread area of the village ponds possessed by the respondent (on lease basis) for practising composite fish culture. It is expressed in hectares. Further, based on the mean and S.D. as measures of check, village tanks are classified into three groups. Actual area (in hectares) has been used for further analyses.

Category	*Scores*	
	North 24 Parganas	*Dakshin Dinajpur*
Low (up to X- ½ SD)	Up to 0.3505	Up to o.292
Medium (X- ½ SD to X+ ½ SD)	0.3505 to 0.7935	o.292 to 0.924
High (above X+ ½ SD)	Above 0.7935	Above 0.924
	Mean : 0.572 SD : 0.443	Mean : 0.608 SD : 0.632

Duration of Water Availability

It refers to the average water retention period in the village ponds in which the respondents are practising composite fish culture. Data pertaining to water retention were collected and categorized into two groups *viz.* short seasonal and long seasonal. The scoring pattern followed for the duration of water availability is shown below:

Category	*Water Retention Period*	*Score*
Short seasonal tanks	6-7 months	1
Long seasonal tanks	8-12 months	2

Source of Water

It refers to the main source of water for village ponds possessed by respondents for practising fish culture. Score given for different water sources of fish culture ponds utilized by the respondents are as follow:

Source of Water	*Score*
Rainfed	1
Canalled	2
Both rainfed and canalled	3

Extent of Weed Infestation

It refers to the presence and the extent of abundance of aquatic weeds in fish culture tanks utilized by the respondents for composite fish culture. Data pertaining to the extent of weed infestation as reported by respondents were collected and categorized into four groups *viz.*, completely chocked, moderate extent, low extent and no weeds. The scoring pattern for measuring the extent of weed infestation is mentioned below.

Category	*Score*
Completely chocked with weeds	1
Moderate extent of weeds	2
Low extent of weeds	3
No weeds	4

Descriptive Variable

Constraints to Adoption of Scientific Fish Culture

A set of constraints or problems were prepared on the basis of an adequate Review of Literature, discussions and preliminary survey in study area. The constraints or problems were classified into eight classes as given below :

A. Problems related to seed, B. Problems related to manure and fertilizers, C. Problems related to artificial feed, D. Problems related to use of pesticides/ insecticides/fish poison, E. Problem related to social situation F. Problem related to marketing, G. Problems related to credit, H. Problems related to transportation.

Although 24 problems were selected from eight areas mentioned above, respondents were requested to check any one of the response categories (level of seriousness) against each problem. The respondents were also asked to describe any other problems faced by them in addition to the above mentioned problems. The responses thus received were rated on a three point continuum as given below :

Category	*Score*
Very serious	2
Serious	1
Not so serious	0

The intensity of each problem was worked out by summing up all the individual scores obtained by the respondents on each of the problems. The problems were then categorized on their total seriousness scores as the respondents perceived. These were categorized into three categories viz, very serious, serious and not so serious by the procedure followed by Sigman and Swanson (1984) as mentioned earlier.

Tools and Techniques of Data Collection

Through a realistic pre-testing a researcher can learn more about all aspects of the planned procedure (Fowler, 1987). In the present study, the entire set of draft interview schedules containing all the above mentioned aspects were subjected to pre-testing with 10 per cent non sample respondents before administering into the actual respondent. In the light of the experience of the pre-testing, the final schedules were revised and modified with appropriate wordings and contents. For easy understanding the schedules were prepared as simple and to the point as possible. In preparation of schedule opinion of the experts were also considered. The final interview schedule was given in Appendix.

Field data were collected with the help of structured and pre-tested interview schedule developed for this purpose. All the data were collected from the respondents through personal interview.

Before interviewing the respondents, they were explained the purpose of conducting the research study and data collection from them. Moreover, they were also assured that the information provided by them will be kept confidential and will only be utilized for writing the research report as a whole. This facilitated to establish the rapport with them as to get the free frank responses and thereby reliable data. Besides necessary information from secondary sources, like census, Government publications, research journal, books, reports and other published documents was also made use of. The data collection was done from March, 2006 to November, 2006.

Statistical Tools Used in the Study

Statistical Analysis : The collected data were complied and tabulated.The data were also categorized for further clarity, realization and interpretation. In the present study, the following methods were used in the analysis of data using the standard formula.

1. Frequency and percentage
2. Correlation test
3. Multiple regression analysis
4. Path analysis

Frequency and Percentage

The data were presented in frequency and percentage to understand the nature of distribution of fish farmers and to know attitude and extent of adoption and constraints prevailing.

Pearson's Correlation Co-efficient

Correlation Co-efficient is a measure of the measure of the relationship between two variables which are at the interval or ratio level of measurement and are linearly related. The range of correlation coefficient (r) lies between -1 to +1. Positive values of 'r' indicate positive correlation between the two variables, *i.e.* changes in both variables take place in the same direction, where as negative values of 'r' indicate negative correlation *i.e.* changes in the two variables falling place in the opposite directions. A zero value of 'r' indicates that there is no linear correlation between the two variables. r=+1 indicates perfect positive correlation and r=-1, it indicates perfect negative correlation, meaning there by that variations in independent variables (x) explain 100 per cent of variations in the dependent variables (y). To get the Pearson correlation coefficient, r, we used *r* SPSS 10.0 software (http://www.esub.edu/ssric-trd/SPSS11-7/11-7htm).

Stepwise Multiple Regression

Stepwise multiple regression is a variation of multiple regression which provides a means of choosing independent variables that yield the best prediction possible with the fewest independent variables. It permits the user to solve a sequence of one or more multiple linear regression problems by stepwise application of the least square method. At each step in the analysis, a variable is added or removed which results in the greatest production in error sum of square. The method of stepwise multiple regression analysis to insert variables in turn until the regression equation is satisfactory. The order of insertion is determined by using the partial correlation coefficient as a measure of the importance of variables not yet in the equation. The programme, according to Burroughs Corporation (1975), first forms a correlation matrix, finds the best predictor (the independent variable having the highest correlation with the criterion variable) and performs a regression analysis with this predictor. Then, the second best predictor (independent variable having the second highest correlation with the criterion) is found and a regression analysis using the multiple correlation of the two best predictors is performed. At any given step, the group of predictors being used is not necessarily the best group of that size. Rather this group contains the variables that have the highest individual correlation with the criterion.

The main objective in using this technique is to predict the variability of the dependent variable based in its covariance with all the independent variables. One can predict the level of the dependent phenomenon through multiple regression analysis model, given the levels of independent variables.

Path Analysis

Path analysis is an extension of the regression model, used to test the fit of the correlation matrix against two or more causal models, which are being compared by the researcher. The model is usually depicted in a circle-and-arrow figure in which single arrows indicate causation. A regression is done for each variable in the model as a dependent on others which the model indicates are causes. The regression weights predicted by the model are compared with the observed correlation matrix for the variables, and a goodness-of-fit statistic is calculated. The best fitting of two or more models is selected by the researcher as the best model for advancement of theory. Path analysis requires the usual assumptions of regression. It is particularly sensitive to model specification because failure to include relevant causal variables or inclusion of extraneous variables often substantially affects the path coefficients, which are used to assess the relative importance of various direct and indirect causal paths to the dependent variable.

Key Concepts of Path : (Wright, S. 1934)

☆ **Path model :** A path model is a diagram relating independent, intermediary, and dependent variables. Single arrows indicate causation between exogenous or intermediary variables and the dependent(s). Arrows also connect the error terms with their respective endogenous variables. Double arrows indicate correlation between pairs of exogenous variables. Sometimes the width of the arrows in the path model are drawn in a width which is proportional to the absolute magnitude of the corresponding path coefficients (see below).

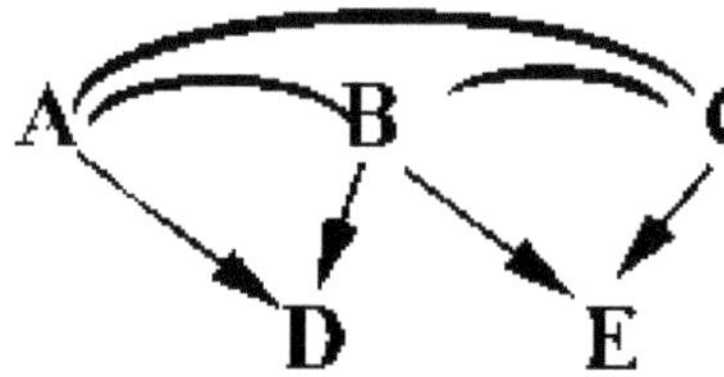

☆ **Causal paths** to a given variable include (1) the direct paths from arrows leading to it, and (2) correlated paths from endogenous variables correlated with others which have arrows leading to the given variable. Consider this model:

This model has correlated exogenous variables A, B, and C, and endogenous variables D and E. Error terms are not shown. The causal paths relevant to variable D are the paths from A to D, from B to D, and the paths reflecting common anteceding causes — the paths from B to A to D, from C to A to D, and from C to B to D. Paths involving two correlations

(C to B to A to D) are not relevant. Likewise, paths that go backward (E to B to D, or E to B to A to D) reflect common effects and are not relevant.

Exogenous and Endogenous Variables

Exogenous variables in a path model are those with no explicit causes (no arrows going to them, other than the measurement error term). If exogenous variables are correlated, this is indicated by a double-headed arrow connecting them. Endogenous variables, then, are those, which do have incoming arrows. Endogenous variables include intervening causal variables and dependents. Intervening endogenous variables have both incoming and outgoing causal arrows in the path diagram. The dependent variable(s) have only incoming arrows.

Path Coefficient/Path Weight

A path coefficient is a standardized regression coefficient (beta) showing the direct effect of an independent variable on a dependent variable in the path model. Thus when the model has two or more causal variables, path coefficients are partial regression coefficients which measure the extent of effect of one variable on another in the path model controlling for other prior variables, using standardized data or a correlation matrix as input. Recall that for bivariate regression, the beta weight (the b coefficient for standardized data) is the same as the correlation coefficient, so for the case of a path model with a variable as a dependent of a single exogenous variable (and an error residual term), the path coefficient in this special case is a zero-order correlation coefficient.

Chapter 5

Findings and Discussion

The results of the study are analysed and discussed to be presented in three broad parts. The first part seeks to present the results yielded by the descriptive analyses of the data relating to the personal, socio-economic, communicational and psychological characteristics of the fish farmers selected for the study. The second part presents the findings resulting from the relational analysis of the data. The third part consists of a discussion of the constraints and problems encountered by the fish farmers in their adoption of scientific fish culture practices, followed by certain suggestions for tackling them.

PART-I

Background Information Relating to the Fish Farmers

In order to understand the factors affecting fish farmers clearly and comprehensively for the purpose of this study, their personal, socio-economic, communicational, psychological and situational characteristics were taken up for close examination, in so far as these related to such aspects as their age, caste, family size, family type, education, fish farming experience, occupation, annual income, land holding, social participation, possession of fishing equipment, mass media participation, extension agency contact, cosmopoliteness, innovative proneness, credit orientation, value orientation, risk orientation, economic motivation, knowledge about scientific fish culture practices, size of water body, duration of water availability, source of water and extent of weed infestation. The manner in which the parameters in question have been dealt with can be seen from the following analyses.

Socio-Personal Variable

1) Age

North 24 Parganas

The age of the fish farmers varies from 19 years to 61 years. Out of 120 fish farmers, the majority, *i.e.* 56 (46.7 per cent) are in the age group of 31-45 years (middle aged), 51 (42.5 per cent) are in the age group of 30 years or below (young aged) and the remaining 13 (10.8 per cent) are in the age group of 46 years and above (Old aged).

Dakshin Dinajpur

The age of the fish farmers varies from 19 years to 61 years Out of 120 fish farmers, the majority, *i.e.* 54 (45.0 per cent) are in the age group of 31-45 years (middle aged), 41 (34.2 per cent) are in the age group of 30 years or below (young aged) and the remaining 25 (20.8 per cent) are in the age group of 46 years and above (Old aged).

Table 3: Distribution of Respondents Based on Age

Category	*North 24 Parganas*		*Dakshin Dinajpur*	
	Frequency	*Percentage*	*Frequency*	*Percentage*
Young (upto 30 years)	51	42.5	41	34.2
Middle (31 to 45 years)	56	46.7	54	45.0
Old (46 years and above)	13	10.8	25	20.8

The results of the study in respect of the two districts are in consonance with the findings of Krishnaiah (1989) and Mahandra Kumar (1996). This parity of results

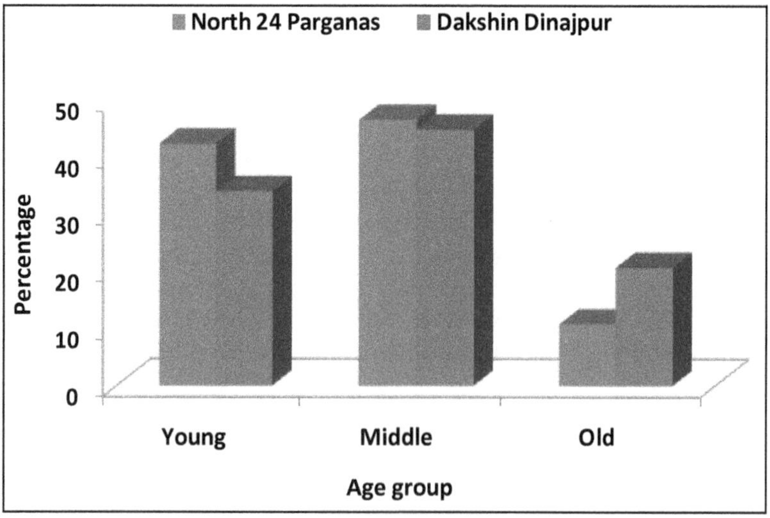

Figure 6: Percentage of Fish Farmers Belonged to different Age Groups.

may be attributed to the adoption in the past of certain measures by the authorities in these districts of generally encouraging the involvement of those fish farmers in qualitative improvement programmes who represent a reasonable mix of youthful enterprise and a modicum of experience and wisdom. It is significant that the FFDA authorities attached due importance to involving in their programme energetic, young and middle aged beneficiaries, who took active part in the adoption of improved fish culture practices.

2) Caste

North 24 Parganas

The results of the study reveal that the majority 74 (61.7 per cent) of the fish farmers belong to general caste whereas 37 (30.8 per cent) fish farmers belong to the scheduled caste category. Only 9 (7.5 per cent) respondents belong to scheduled tribe.

Dakshin Dinajpur

The results show that the majority 66 (55.0 per cent) of the fish farmers belong to general caste whereas 41 (34.2 per cent) fish farmers belong to the scheduled caste category. Only 13 (10.8 per cent) respondents belong to scheduled tribe.

Table 4: Distribution of Respondents Based on their Caste

Category	*North 24 Parganas*		*Dakshin Dinajpur*	
	Frequency	*Percentage*	*Frequency*	*Percentage*
General	74	61.7	66	55.0
SC	37	30.8	41	34.2
ST	9	7.5	13	10.8

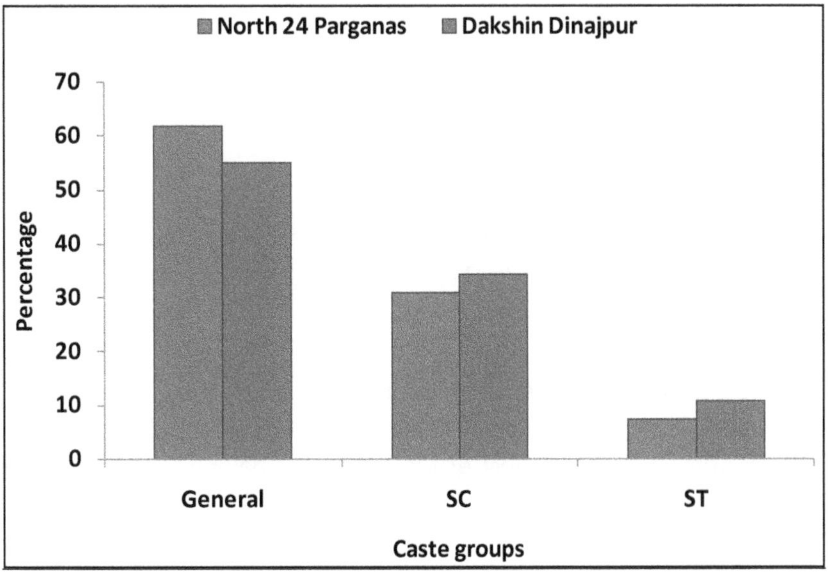

Figure 7: Distribution of Fish Farmers Based on their Caste.

The above results are not in agreement with the findings of several earlier surveys (Subhas Chandra, 1986; Munda *et al.*, 1997; Singh and Nanda, 2001 and Ghosh *et al.*, 2002). According to those surveys, the majority of the fish farmers under study belonged to SC, ST and/or other backward classes. The departure of the findings of the present study may be attributed to the operation of such differential factors as higher income, higher social participation, large size of ponds, exposure to training and the like.

3) Family Size

North 24 Parganas

The results show that both the small family and large family categories have an almost an equal share amongst the total size of the respondents.

Dakshin Dinajpur

The results reveal that the respondents belonging to the small family and the large family categories are almost equally represented amongst the respondents.

Table 5: Distribution of Respondents According to their Family Size

Category	*North 24 Parganas*		*Dakshin Dinajpur*	
	Frequency	*Percentage*	*Frequency*	*Percentage*
Small family(upto 5 members)	61	50.8	59	49.2
Large family(More than 5 members)	59	49.2	61	50.8

The analysis of the family size of the respondents (table-5) shows almost equal distribution of the respondents between the small family unit (upto 5 members) and large family unit (size more than five). Incidentally, the techno socio-economic survey of fishermen community in earlier studies revealed that the average size of household of inland fishermen in Karnataka was 5.7 (Nageswara Rao *et al.*, 1998) and in Lake Jaisamand it was about 5 members (Sharma and Kumar, 2001).

4) Family Type

North 24 Parganas

With regard to the family type of fish farmers, 68 respondents (56.7 per cent) are found to come from a nuclear type of family and 52 respondents (43.3 per cent) from joint family.

Dakshin Dinajpur

The results reveal that 64 respondents (55 per cent) are from nuclear type of family and 56 respondents (46.7 per cent) are from joint family.

The results reveal that most of the fishermen belong to nuclear family (55 per cent), followed by joint family (45 per cent). Similar findings are reported by Govinda Gowda *et al.* (1991).

Table 6: Distribution of Respondents Based on their Family Type

Category	*North 24 Parganas*		*Dakshin Dinajpur*	
	Frequency	*Percentage*	*Frequency*	*Percentage*
Nuclear	68	56.7	64	53.3
Joint	52	43.3	56	46.7

5) Education

North 24 Parganas

Out of 120 fish farmers, the majority, *i.e.*, 108 (90.0 per cent) are literate and the remaining 12 (10.0 per cent) are illiterate. Among the literate fish farmers the majority, *i.e.*, 35 (29.2 per cent) have a level of education upto middle school followed by 30 (25.0 per cent) fish farmers with a primary school level of education. There are 18 (15.0 per cent) fish farmers who can only read and write in Bengali, 13 (10.8 per cent) fish farmers educated up to the secondary level while, there are only 12 (10 per cent) fish farmers who have higher secondary level of education.

Dakshin Dinajpur

The results show that majority *i.e.*, 102 (85.0 per cent) are literate and the remaining 18 (15.0 per cent) are illiterate. Among the literate fish farmers majority, *i.e.*, 39 (32.5 per cent) had primary school level education followed by 33 (27.5 per

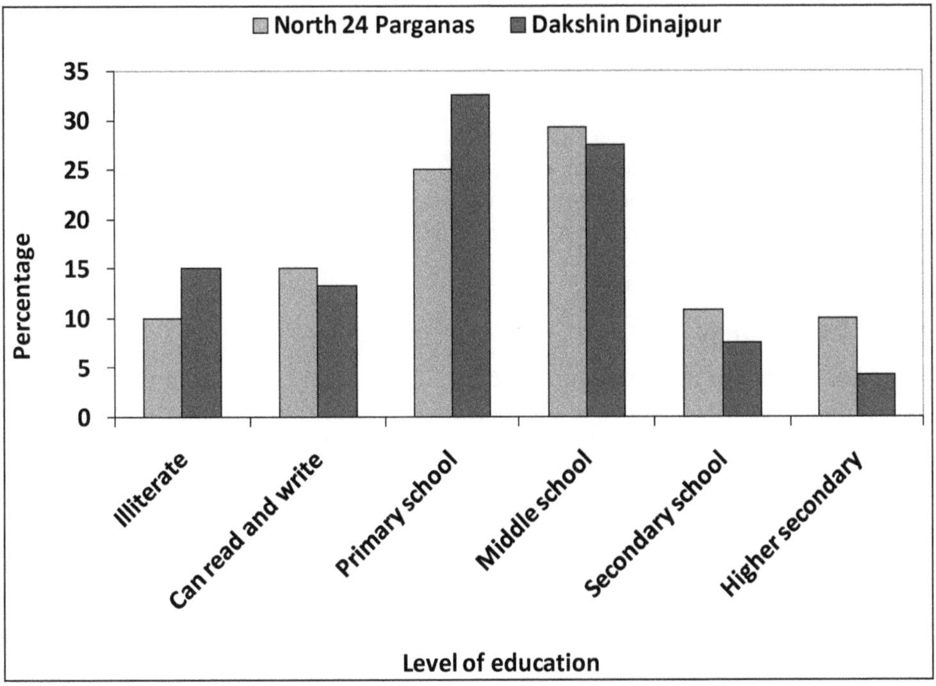

Figure 8: Distribution of Respondents Based on their Education.

cent) fish farmers with a middle school level of education. There are 16 (13.3 per cent) fish farmers who could only read and write in Bengali, 9 (7.5 per cent) fish farmers with a secondary level of education and, interestingly, there are only 5 (4.2 per cent) fish farmers who receive a graduate level of education.

Table 7: Distribution of Respondents According to their Education

Category	*North 24 Parganas*		*Dakshin Dinajpur*	
	Frequency	*Percentage*	*Frequency*	*Percentage*
Illiterate	12	10.0	18	15.0
Can read and write	18	15.0	16	13.3
Primary school	30	25.0	39	32.5
Middle school	35	29.2	33	27.5
Secondary school	13	10.8	9	7.5
Higher secondary	12	10.0	5	4.2

It can be observed from the results of the educational status of the fish farmers in the two districts that the overall level of academic education is far from satisfactory and a few of the fish farmers are even downright illiterate while only a few have privileged to acquire education beyond high school level. This incidentally goes to show that the opportunities available for education upto high school level are well within the reach of the villagers. However, education of this naturally scarcely guarantees employment. They are generally found unemployed after finishing schooling. Ultimately they engage in fish farming as a profession by way of self employment. Similar results are reported by Pounraj and Sripal (1997), Mahandra Kumar (1996), Rajukumar (1988), Meeran (1983), Balasubramaniam (1988), Govinda Gowda *et al.* (1991), Wayal and Ingle (1994), Manker *et al.* (2000). The instant results of the present study revealing that farmers belong to a low level of education have practical implications. The curriculum in respect of scientific fish culture to be evolved for them need to be prepared according to the level of their understanding and comprehension and in consonance with their level of education.

6) Fish Farming Experience

North 24 Parganas

Majority of the respondents, *i.e.*, 62 (51.7 per cent) belong to a medium level of experience in fish farming while 51 (42.5 per cent) of them have a low level of experience, only 7 (5.8 per cent) possess high level of experience (more than 5 years) in fish farming.

Dakshin Dinajpur

The results reveal that majority of the fish farmers *i.e.*, 59 (49.2 per cent) belong to a low level of experience (upto 2 years) while 50 (41.6 per cent) of them have a medium level of experience and only 11 (9.2 per cent) have a high level of experience (more than 5 years).

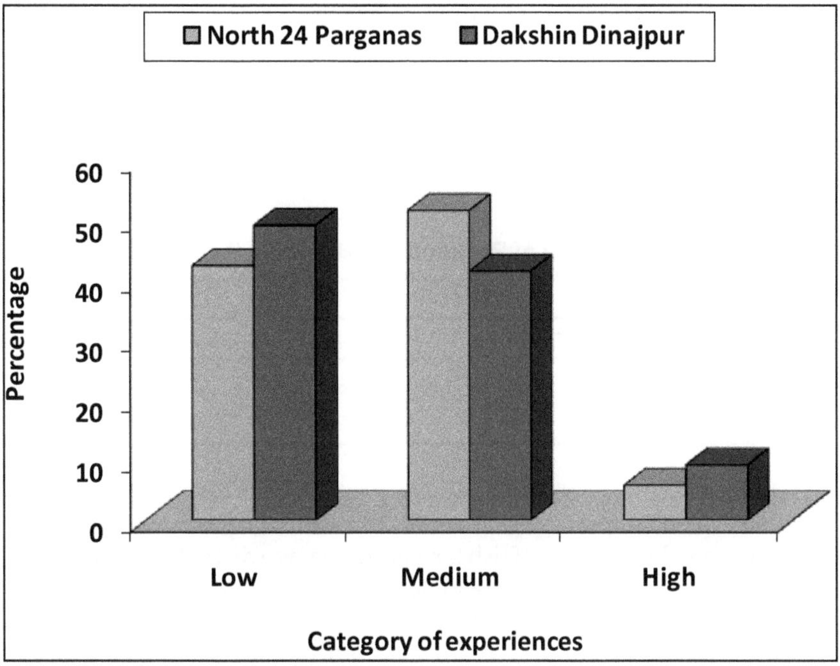

Figure 9: Distribution of Respondents Based on their Fish Farming Experiences.

Table 8: Distribution of Respondents According to Experience in Fish Farming

Category	*North 24 Parganas*		*Dakshin Dinajpur*	
	Frequency	*Percentage*	*Frequency*	*Percentage*
Low (Up to 2 years)	51	42.5	59	49.2
Medium (3 to 6 years)	62	51.7	50	41.6
High (7 years and above)	7	5.8	11	9.2

The number of fish farmers with a high level of experience (7 yrs. and above) is negligible in both the districts. In earlier days under FFDA programme, ponds were leased out to beneficiaries for a period of 5-10 years but of late the period of leasing has been reduced to a reasonable term of 5-6 years. The above table demonstrates the variations in fish farming experiences among fish farmers and tends to build up a pattern obtaining representatively in the rural areas of West Bengal where there is a considerable scope of fish farming. The results are in agreement with the findings of earlier studies (Ponnappan, 1982 and Krishnaiah, 1989).

Socio-economic Variables

7) Occupation

North 24 Parganas

The data indicate that for the majority of the fish farmers, *i.e.*, 89 (74.2 per cent) fish farming is a secondary occupation as against 31 (25.8 per cent) respondents taking it up as the main occupation.

Dakshin Dinajpur

The results reveal that for the majority of the respondents, *i.e.*, 93 (77.5 per cent) fish farming is a secondary occupation while it is the main occupation for 27 (22.5 per cent) respondents.

Table 9: Distribution of Respondents Based on their Occupation

Category	*North 24 Parganas*		*Dakshin Dinajpur*	
	Frequency	*Percentage*	*Frequency*	*Percentage*
Main occupation	31	25.8	27	22.5
Secondary occupation	89	74.2	93	77.5

The trend or pattern seems to be clearly established. Fish farming is generally found to have been adopted by farmers as their secondary occupation. The present study reveals that the majority of fish farmers of the two districts took to fish farming as a secondary occupation. This general tendency has its rationale. In all probability the unemployed youth from agricultural and non-agricultural families were trained and asked to opt for fish culture to earn some additional income since they already possessed lands for cultivation which they could fall back upon in any way. Similar findings are reported by Meeran (1983), Subhas Chandra (1986), Mahandra Kumar (1996), Krishnaiah (1989), Govinda Gowda *et al.* (1991) and Singh and Nanda (2001).

8) Annual Income

North 24 Parganas

This study reveals that the majority *i.e.* 82 (68.3 per cent) are from higher income group (Rs. Above 0.33 lakh) as against which 19 (15.8 per cent) are from low medium annual income group (Rs. 0.11 lakh to Rs 0.22 lakh) and 11 (9.2 per cent) are from medium annual income group (Rs. 0.22 lakh to Rs 0.33lakh) while the remaining 8 (6.7 per cent) fish farmers are from low income group (Up to Rs. 0.11 lakh). The average annual income of the fish farmers is Rs. 0.65 lakh. The minimum annual income of the fish farmers is Rs. 0.08 lakh and the maximum is Rs 2.4 lakh.

Dakshin Dinajpur

The data indicate that the majority *i.e.* 78 (65.0 per cent) are from higher income group (Rs. Above 0.33 lakh) whereas 22 (18.33 per cent) are from low medium annual income group (Rs. 0.11 lakh to Rs 0.22 lakh) and 10 (8.33 per cent) are from medium annual income group (Rs. 0.22 lakh to Rs 0.33lakh) and the remaining 8 (6.7 per cent) fish farmers are from low income group (Up to Rs. 0.11 lakh). The average annual income of the fish farmers is Rs. 0.477 lakh. The minimum annual income of the fish farmers is Rs. 0.05 lakh and the maximum is Rs 1.0 lakh.

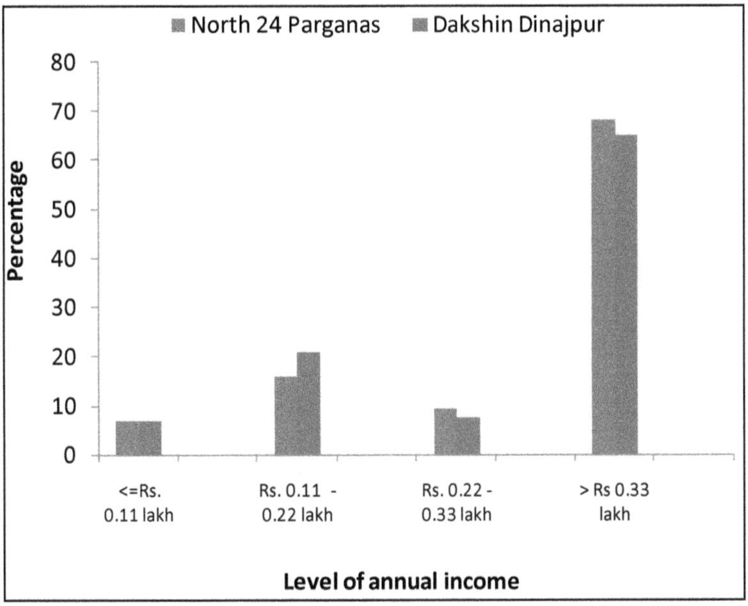

Figure 10: Distribution of Respondents Based on their Annual Income.

Table 10: Distribution of Respondents According to their Annual Income

Category	*North 24 Parganas*		*Dakshin Dinajpur*	
	Frequency	*Percentage*	*Frequency*	*Percentage*
Up to Rs. 0.11 lakh	8	6.7	8	6.7
Rs. 0.11 lakh to Rs 0.22 lakh	19	15.8	25	20.8
Rs. 0.22 lakh to Rs 0.33lakh	11	9.2	9	7.5
Above Rs 0.33 lakh	82	68.3	78	65.0

A good number of farmers of the two districts are under high income categories (above Rs. 0.33 lakh). The findings of the study are in line with the studies of Pounraj (1992) and Pounraj and Sripal (1997).

9) Land Holding

North 24 Parganas

The data indicate that the majority of the respondents, *i.e.*, 69 (57.5 per cent) possess small holdings (Up to 2 acres) of land whereas 29 respondents belong to small to medium holding category (2 to 5 acres) and 22 farmers belong to medium to large holding category (above 5 acres each).

Dakshin Dinajpur

This study reveals that most of the fish farmers, *i.e.*, 59 (49.2 per cent) of the respondents possess small holdings (Up to 2 acres each) of land whereas 50

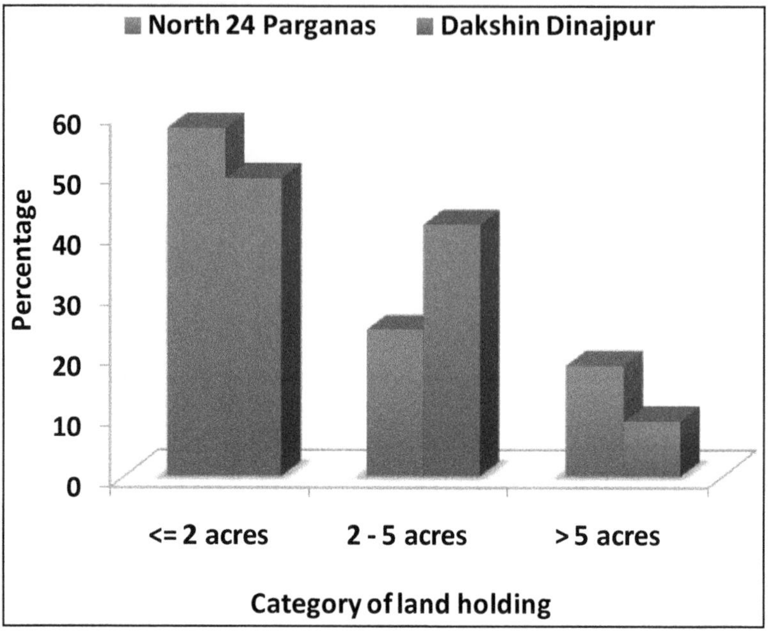

Figure 11: Distribution of Respondents Based on their Land Holding.

respondents belong to small to medium holding category (2 to 5 acres) and 11 farmers belong to medium to large holding category (above 5 acres).

Table 11: Distribution of Respondents According to Land Holding

Category	*North 24 Parganas*		*Dakshin Dinajpur*	
	Frequency	*Percentage*	*Frequency*	*Percentage*
Up to 2 acres	69	57.5	59	49.2
2 to 5 acres	29	24.2	50	41.6
More than 5 acres	22	18.3	11	9.2

The results indicate that the majority of the fish farmers of the two districts possess low land holding (upto 2 acres). Mahandra Kumar (1996) reported medium land holding by 59 per cent and small land holding by 33 per cent of fish farmers in Tamil Nadu. Ponnappan (1982) reported that about 51 per cent of fish farmers were small and marginal farmers. Wayal and Ingle (1994) observed that most of the fish farmers are landless (69.39 per cent) followed by marginal farmers (13.26 per cent).

10) Social Participation

North 24 Parganas

The data pertaining to the member/office bearer and participation of the respondents in various social organizations and local institutions reveal that the majority 70 (58.3 per cent) belong to medium social participation category followed

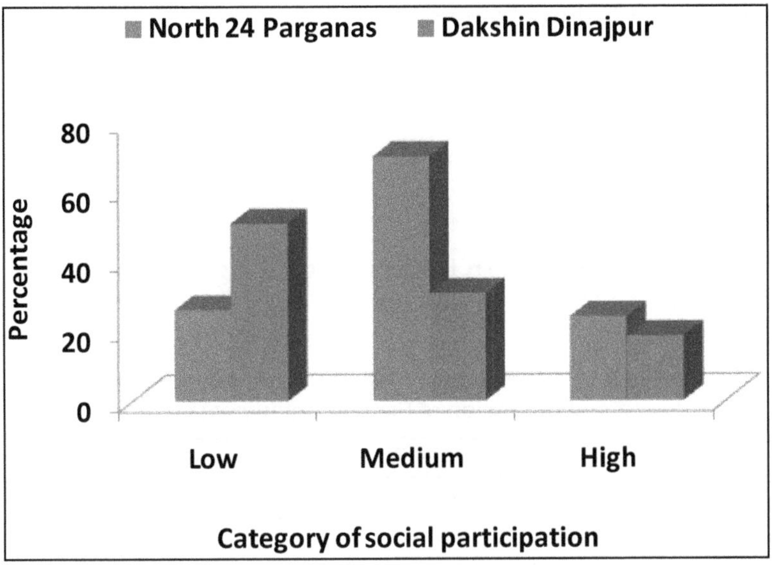

Figure 12: Distribution of Respondents Based on their Social Participation.

by 26 (21.7 per cent) of the respondents belonging to low social participation, while the remaining 24 (20 per cent) respondents belong to high social participation category.

Dakshin Dinajpur

The results reveal that the majority, *i.e.*, 61 (50.8 per cent) of the respondents belong to low social participation category followed by 37 (30.8 per cent) of the respondents belonging to medium social participation, while the remaining 22 (18.4 per cent) respondents belong to high social participation category.

Table 12: Distribution of Respondents Based on their Social Participation

Category	*North 24 Parganas*		*Dakshin Dinajpur*	
	Frequency	*Percentage*	*Frequency*	*Percentage*
Low social participation	26	21.7	61	50.8
Medium social participation	70	58.3	37	30.8
High social participation	24	20	22	18.4

The research findings reveal that in the case of the majority of the fish farmers in Dakshin Dinajpur district the level of social participation is low. In case of the majority of the fish farmers of North 24 Parganas district, however, the level of social participation is medium.

The earlier studies by Subhas Chandra (1986) and Krishnaiah (1989) reported medium to low levels of social participation of fish farmers in Andhra Pradesh. In Dakshin Dinajpur the low social participation can be attributed to factors like poor education, low socio-economic status and low mass media participation.

11) Possession of Fishing Equipments

North 24 Parganas

The results reveal that the majority, *i.e.*, 73 (60.8 per cent) of the respondents have fishing equipments of their own while only 39.2 per cent of the fish farmers have no fishing equipments.

Dakshin Dinajpur

A greater percentage of fish farmers have fishing equipments of their own 69 (57.5 per cent) while a meagre 42.5 per cent of the fish farmers have no fishing equipments.

Table 13: Distribution of Respondents According to Possession of Fishing Equipments

Category	*North 24 Parganas*		*Dakshin Dinajpur*	
	Frequency	*Percentage*	*Frequency*	*Percentage*
No fishing equipment	47	39.2	51	42.5
Having fishing equipments	73	60.8	69	57.5

results reveal that most of the fish farmers of the two districts possess fishing equipments of their own. The findings of earlier studies (Krishnaiah 1989, Ponnappan, 1982, Subhas Chandra, 1986 and Pounraj and Sripal, 1997) indicated medium level of possession of fishing equipments by the respondents. The present findings are more or less in conformity with the previous findings in question.

Communication Variables

12) Mass Media Participation

North 24 Parganas

The majority of fish farmers *i.e.*, 62 (51.7 per cent) belong to the category of medium level of mass media participation, followed by 32 (26.7 per cent) respondents evincing a low level of participation whereas only 21.6 per cent of the fish farmers come under the category of high level of media participation.

Table 14: Distribution of Respondents Based on their Mass Media Participation

Category	*North 24 Parganas*		*Dakshin Dinajpur*	
	Frequency	*Percentage*	*Frequency*	*Percentage*
Low	32	26.7	54	45.0
Medium	62	51.7	34	28.3
High	26	21.6	32	26.7

Dakshin Dinajpur

This study reveals that most of the fish farmers *i.e.* 54 (45.0 per cent) belong

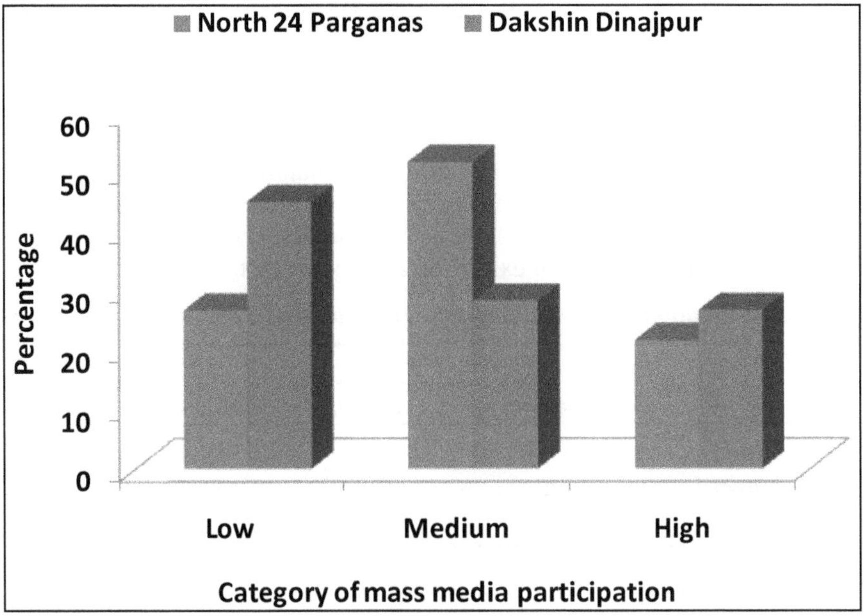

Figure 13: Distribution of Respondents Based on their Mass Media Participation.

to the low level of mass media participation, followed by medium level (28.3 per cent) participation whereas only 26.7 per cent of the fish farmers come under the category of high level of media participation.

Thus so far as this parameter is concerned, we notice a considerable disparity between the fish farmers of the two districts under study. While in case of North 24 Parganas, the predominant trend as regards the level of mass media participation is medium; in the case of Dakshin Dinajpur it is rather low, indicating a very poor level of participation and utilization of the benefit of different media participation for acquiring technological information by the fish farmers. The situation in the latter district, however, can and should be improved by taking appropriate remedial measures. Mass media sources are at present being utilized only to a limited extent, but it is still possible to improve the level of utilization by frequent and effective use and also by implementing other propaganda techniques by fisheries extension personnel. Constant and repetitive use of many sources of information could instil confidence into them and persuade them to adopt scientific fish culture practices. This will lead to better coverage for the purpose. Improved and heightened dissemination of extension messages through mass media like TV, radio, newspaper, agriculture magazines could ensure coverage of more farmers in a relatively short time and at negligible additional cost. The findings are in line with the results obtained by Wayal (1990), Govinda Gowda *et al.* (1991) and Sushant *et al.* (2002).

13) Extension Agency Contact

North 24 Parganas

A large section of fish farmers *i.e.,* 52 (43.3 per cent) have a medium level of extension agency contact followed by 44 (36.7 per cent) with a low level of extension agency contact. There are 24 (20.0 per cent) fish farmers with a high level of extension agency contact.

Dakshin Dinajpur

With regard to extension agency contact, the majority, *i.e.*, 51 (42.5 per cent) of fish farmers have a medium level of extension agency contact followed by 46 (38.3 per cent) with a low level of extension agency contact. There are 23 (19.2 per cent) fish farmers with a high level of extension agency contact.

Table 15: Distribution of Respondents According to their Extension Agency Contact

Category	*North 24 Parganas*		*Dakshin Dinajpur*	
	Frequency	*Percentage*	Frequency	Percentage
Low	44	36.7	46	38.3
Medium	52	43.3	51	42.5
High	24	20.0	23	19.2

It is found that the majority of the fish farmers fall under the category of medium to low level of extension agency contact. This is perhaps attributable to poor extension activities by the State Fishery Department. In most cases respondents are inclined to make contact with the Fishery Extension Officers' their residences being close to the FEO office. But in reality such contacts scarcely materialize as they occasionally get the information from the Fishery Extension Officer.

Fish farmers also receive information about scientific fish culture practices from other extension agencies like Krishi Vigyan Kendra, progressive farmers input dealers. The findings are in conformity with the studies undertaken by Meeran (1983), Pounraj (1992), Pounraj and Sripal (1997), Sushant *et al.* (2002).

14) Cosmopoliteness

North 24 Parganas

The majority of fish farmers, *i.e.*, 51 (42.5 per cent) belong to the category of high cosmopoliteness followed by a medium level (35.8 per cent) of the kind. Almost 1/5th of the fish farmers (21.7 per cent) come under low cosmopoliteness category.

Table 16: Distribution of Respondents Based on their Cosmopoliteness

Category	*North 24 Parganas*		*Dakshin Dinajpur*	
	Frequency	*Percentage*	*Frequency*	*Percentage*
Low	26	21.7	43	35.8
Medium	43	35.8	39	32.5
High	51	42.5	38	31.7

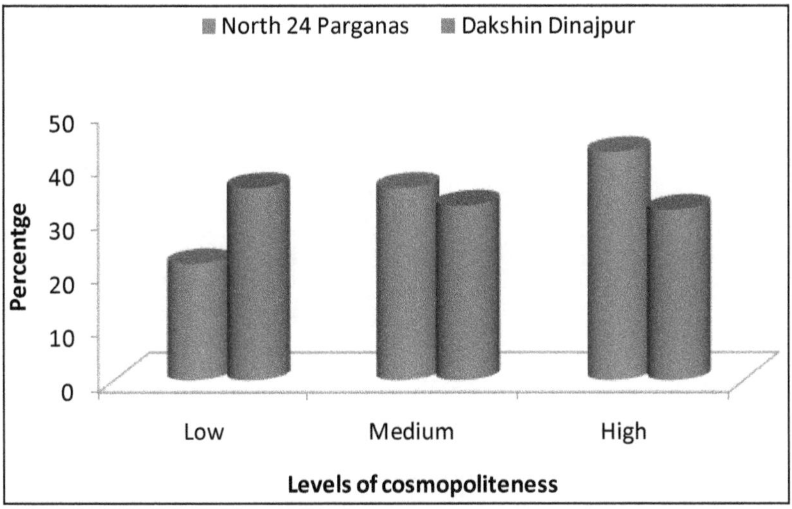

Figure 14: Distribution of Respondents Based on their Cosmopoliteness.

Dakshin Dinajpur

This study reveals that most of the fish farmers, *i.e.*, 43 (35.8 per cent) of the total number belong to the category of low cosmopoliteness followed by the medium level cosmopolite group (32.5 per cent). whereas only 31.7 per cent of fish farmers come under the high cosmopoliteness category.

It is clear from the results that the majority of the fish farmers are grouped under medium to high category of cosmopoliteness in North 24 Parganas whereas in Dakshin Dinajpur district the majority of the fish farmers are grouped under low to medium categories of cosmopoliteness. The preponderance of this factor has a critical effect. Cosmopoliteness prepares the individuals to acquire more knowledge from different sources and provides an opportunity to them for better exchange of ideas and information. The findings from North 24 Parganas are nearly in conformity with the details reported by Talukdar (2000).

Psychological Variables

15) Innovative Proneness

North 24 Parganas

The results indicate that 49 (40.8 per cent) of the fish farmers have a high level of innovative proneness, followed by lesser sections of them having a medium (34.2 per cent) and a low (25.0 per cent) level of innovative proneness

Dakshin Dinajpur

The data indicate that 38.4 per cent of fish farmers have a high level of innovative proneness, followed by low (35.8 per cent) and medium (25.8 per cent) levels of innovative proneness.

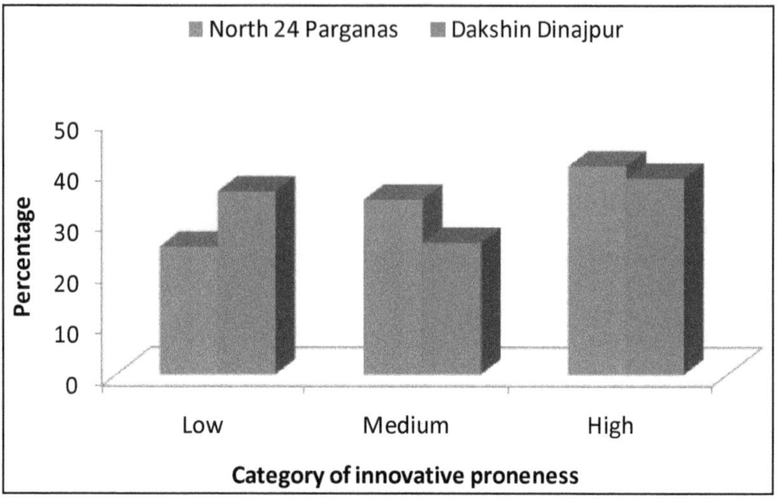

Figure 15: Percentage of Fish Farmers Belonged to different Innovative Proneness Category.

Table 17: Distribution of Respondents According to Innovative Proneness

Category	*North 24 Parganas*		*Dakshin Dinajpur*	
	Frequency	*Percentage*	*Frequency*	*Percentage*
Low	30	25.0	43	35.8
Medium	41	34.2	31	25.8
High	49	40.8	46	38.4

In North 24 Parganas three fourth (75 per cent) of the fish farmers have recorded high to medium level of innovative proneness but in case of Dakshin Dinajpur the majority of the fish farmers belong to high level of innovative proneness followed by those evincing a low innovative proneness. Innovative proneness refers to the behavioural aspects of a fish farmer so far as these relate to or demonstrate a tendency to seek change in fish farming techniques and to introduce such changes into his fish culture practices which are practical and feasible. The higher the innovative proneness the greater the performance acceleration, because higher innovative proneness results in more rapid changes. As such this characteristic of a fish farmer is of supreme importance from the point of view of the ends of the present study. It is found that one third of the fish farmers belong to this high category. Ponnappan (1982) and Pounraj and Sripal (1997) observed a medium level of innovativeness among the majority of the fish farmers brought under their respective studies.

16) Credit Orientation

North 24 Parganas

This study reveals that most of the respondents, numbering 47 (39.2 per cent), belong to low credit orientation, followed by 38.3 per cent of the respondents belonging to medium credit orientation category. It is observed that 27 of the respondents (22.5 per cent) belong to high credit orientation category.

Dakshin Dinajpur

The result indicates that 51 (42.5 per cent) of respondents belong to the category of low credit orientation, followed by 38 per cent of the respondents belonging to the medium credit orientation category. There are 31 (25.8 per cent) fish farmers who belong to the high credit orientation category.

Table 18: Distribution of Respondents According to their Credit Orientation

Category	*North 24 Parganas*		*Dakshin Dinajpur*	
	Frequency	*Percentage*	*Frequency*	*Percentage*
Low	47	39.2	51	42.5
Medium	46	38.3	38	31.7
High	27	22.5	31	25.8

The above findings of the study in both the districts indicate that the majority of the fish farmers have a low to medium category of credit orientation and nearly a quarter of the fish farmers have a high credit orientation. In the present study most of the fish farmers belong to in higher income groups for which they generally are not in need of any financial help for procuring necessary input and sustenance for life during the emergency period. Sushant *et al.* (2002) observed a high credit orientation on the part of the fish farmers.

17) Value Orientation

North 24 Parganas

With regard to value orientation, the majority of fish farmers, *i.e.*, 57 (47.5 per cent) enjoy a medium value orientation while the rest enjoy either a low (28.3 per cent) or a high (24.2 per cent) value orientation.

Dakshin Dinajpur

The data indicate that the majority of fish farmers, *i.e.*, 46 (38.3 per cent) from the high value orientation segment followed by medium (34.2 per cent) and low (27.5 per cent) value orientation segment.

Table 19: Distribution of Respondents According to Value Orientation

Category	*North 24 Parganas*		*Dakshin Dinajpur*	
	Frequency	*Percentage*	*Frequency*	*Percentage*
Low	34	28.3	33	27.5
Medium	57	47.5	41	34.2
High	29	24.2	46	38.3

The findings reveal that in North 24 Parganas district the majority of the fish farmers (75.8 per cent) belong to the category of medium to low level of value orientation. In Dakshin Dinajpur the majority of the fish farmers belong to high to medium level of value orientation. The value orientation of fish farmers is regarded as an important aspect of the personality of the fish farmers. Dana (1987) observed the existence of medium to low level of value orientation amongst the livestock owners.

18) Risk Orientation

North 24 Parganas

The majority of respondents, *i.e.*, 65 (54.2 per cent) have a medium risk orientation followed by those having a low (25.0 per cent) risk orientation, while the remaining 20.8 per cent of the respondents belong to the high risk orientation category.

Dakshin Dinajpur

As many as 63, representing 52.5 per cent of all respondents, forming the majority of them, have a medium risk orientation followed by the low risk orientation (28.3 per cent) group. Only 19.2 per cent of the respondents belong to the high risk orientation category.

Table 20: Distribution of Respondents Based on their Risk Orientation

Category	*North 24 Parganas*		*Dakshin Dinajpur*	
	Frequency	*Percentage*	*Frequency*	*Percentage*
Low	30	25.0	34	28.3
Medium	65	54.2	63	52.5
High	25	20.8	23	19.2

The results indicate a medium level of risk orientation for the majority of the fish farmers in both the districts. It could be interpreted as the medium to low level of ability of most of the farmers to invite risk. Similar findings are reported by Chede (1988), Krishnaiah (1989), Mankar *et al.* (2000). The above findings support the fact that scientific fish farming is a risk oriented enterprise and that the readiness to adopt practices involved in it largely depends on the level and strength of the risk orientation of the fish farmers.

19) Economic Motivation

North 24 Parganas

As regards economic motivation, it can be seen that the majority of fish farmers, *i.e.*, 51 (42.5 per cent) come under the medium level of economic motivation, followed by the resting sections having a high (30.8 per cent) and a low (26.7 per cent) economic motivation respectively.

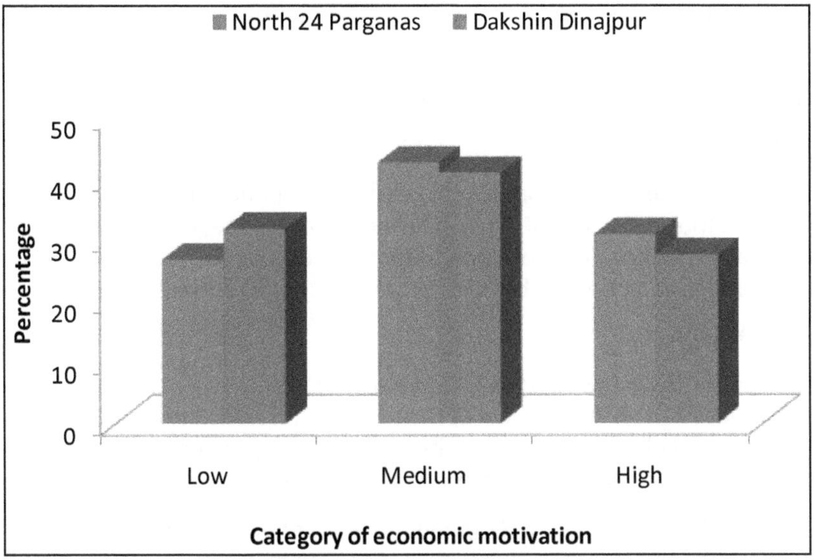

Figure 16: Percentage of Fish Farmers under different Economic Motivation Category.

Dakshin Dinajpur

As regards economic motivation, it is seen the majority of the fish farmers, *i.e.*, 49 (40.8 per cent) have a medium level of economic motivation, followed by those with a low (31.7 per cent) and a high (27.5 per cent) economic motivation.

Table 21: Distribution of Respondents According to Economic Motivation

Category	*North 24 Parganas*		*Dakshin Dinajpur*	
	Frequency	*Percentage*	*Frequency*	*Percentage*
Low	32	26.7	38	31.7
Medium	51	42.5	49	40.8
High	37	30.8	33	27.5

The findings of the study indicate that medium to high level of economic motivation is shared by the majority of the fish farmers in North 24 Parganas whereas in Dakshin Dinajpur the majority of the fish farmers reveal medium to low level of economic motivation. The existence of medium to low level of economic motivation in Dakshin Dinajpur may be attributed to the fact that the fish farmers have failed to receive proper guidance and encouragement from and appraisal by extension officials. This indicates the imperative need for the extension worker properly to disseminate the scientific culture practices with an outline of economic benefit : cost ratio of the technology involved method to the farmers. Similar findings have been reported by Chede (1988), Krishnaiah (1989) and Talukdar (2000).

20) Knowledge Level of Fish Farmers Regarding Scientific Fish Culture Practices

a) Overall Knowledge Level of Fish Farmers

North 24 Parganas

As regards the knowledge level of fish farmers about scientific fish culture practices, it is found that out of 120 fish farmers most of the fish farmers, *i.e.*, 52 (43.3 per cent) have medium level of knowledge whereas 38 (31.7 per cent) have a high level of knowledge, followed by 30 (25.0 per cent) with a low level of knowledge.

Dakshin Dinajpur

As regards their knowledge level about scientific fish culture practices, it is found that out of 120 fish farmers the majority of fish farmers, *i.e.*, 52(43.3 per cent) have medium level of knowledge whereas 38 (31.7 per cent) have high level of knowledge, followed by 30 (25.0 per cent) with a low level of knowledge.

Table 22: Distribution of Respondents According to Knowledge Level

Category	*North 24 Parganas*		*Dakshin Dinajpur*	
	Frequency	*Percentage*	*Frequency*	*Percentage*
Low	30	25.0	30	25.0
Medium	52	43.3	52	43.3
High	38	31.7	38	31.7

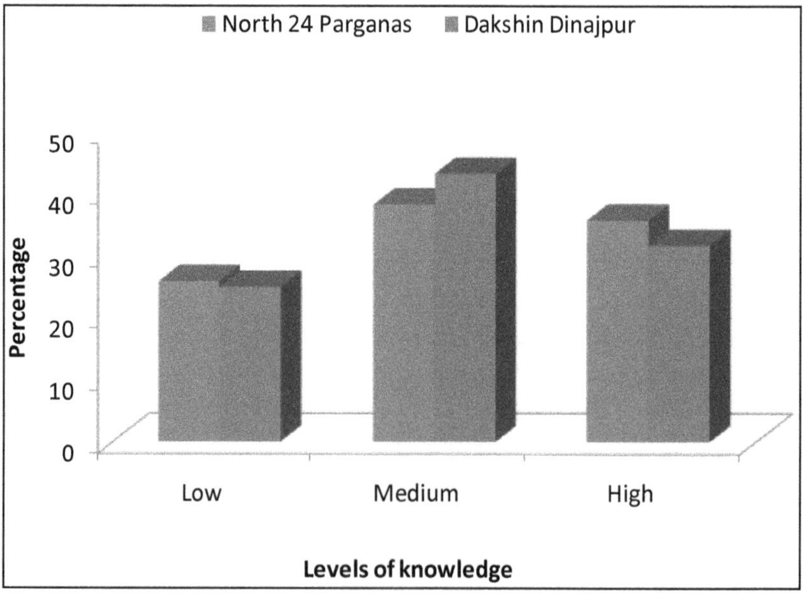

Figure 17: Percentage of Fish Farmers Belonged to different Levels of Knowledge.

The data reveal that the majority of the fish farmers have medium to high level of knowledge regarding scientific fish culture practices in both the districts. Scientific knowledge of any technology is an essential prerequisite for increased productivity. The present findings are in conformity with the findings of Meeran (1983), Misra (1987), Sheela (1991), Beerannavar (1995), Mahandra Kumar (1996), Shinde *et al.* (1998), Meti (1998) and Awasthi *et al.* (2000).

b) Knowledge Level of Fish Farmers Regarding Specific Aspects of Practices Recommended for Scientific Fish Culture

North 24 Parganas

The table-23 indicates that a more or less high percentage of fish farmers have precise knowledge of scientific fish farming in such vital matters or item as the names of some aquatic weeds (95.83 per cent), the need for the use of common organic manures in fish culture (95.83 per cent), the necessity of liming of fish ponds (95 per cent), the need for manuring in the fish ponds (95 per cent), the names of Indian Major Carps (95 per cent), the nature of the soil suitable for fish culture (94.17 per cent), the need for the control of predatory and weed fish (94.17 per cent), the necessity for eradicating excess aquatic weeds (92.5 per cent), names of predatory and weed fishes (90.83 per cent), the indicators of oxygen depletion in the pond (87.5 per cent), the necessity of supplementary feeds (86.67 per cent), the names of fish diseases (86.67 per cent), the manuring schedule recommended to be practised after stocking (85 per cent), knowledge about the remedies for acidic condition of pond water (81.67 per cent), the correct dose of lime (77.5 per cent), the dose of organic manure to be applied (80.0 per cent), the correct species combination in composite fish culture (77.5 per cent) and water quality management (involving necessity to stop manuring and feeding beyond a certain level) (72.5 per cent).

Further, the level of correct knowledge of fish farmers in general, in respect of different items of fish farming varies from item to item, though ranging from slightly above 50 per cent to nearly seventy per cent. For instance, the level in question varies as follows for certain items: advantages of manuring (68.33 per cent), feeding method (67.5 per cent), disease control method (65 per cent), names of the exotic carps (62.5 per cent), fastest growing major carp and exotic carps (59.17 per cent), ideal size of fish seed for stocking (55.83 per cent), the optimum size of harvesting (55.83 per cent), need for the use of inorganic fertilizer (54.17 per cent) and method of eradication of predatory and weed fishes (54.17 per cent).

However, it is clear from table-23 that the majority of the fish farmers lack correct information about such important matters as the knowledge items pertaining to nutrients required for production of fish food organism, the names of fish growing well in weed infested ponds, the advantages of using inorganic fertilizers, the names of piscicides, application of mohua oil cake, harvesting period, the minimum depth of water required for fish culture, the time gap required between manure application and stocking of fish seed, the correct feeding rate of supplementary feed and the necessity of checking growth.

Dakshin Dinajpur

The Table 23 indicates that a more or less high percentage of fish farmers have precise knowledge of scientific fish farming in such vital items as correct knowledge regarding suitable soil for fish culture (85.8 per cent), necessity of liming of fish ponds (90 per cent), need for manuring in fish ponds (90.8 per cent), the need for use of common organic manures in fish culture (86.7 per cent), the names of Indian Major Carps (85 per cent), the necessity of eradicating excess aquatic weeds (78.3 per cent), the names of some aquatic weeds (75.8 per cent), the need for control of predatory and weed fish (80.8 per cent), names of predatory and weed fishes (75.8 per cent), the name of fish grows well in weed infested ponds (77.5 per cent) and the necessity of supplementary feeds (77.5 per cent).

Table 23: Knowledge Level of Fish Farmers Regarding Specific Items of Recommended Scientific Fish Culture (N=240)

Sl.No.	*Practices*	*Response Category (Yes)*			
		North 24 Parganas		*Dakshin Dinajpur*	
		Freque-ncy	*Percen-tage*	*Freque-ncy*	*Percen-tage*
1.	What kind of soil is good for fish culture ?	113	94.17	103	85.83
2.	What is the minimum depth of water required for fish culture ?	32	26.67	24	20
3.	What are the nutrients required for production of natural fish food organisms in fish pond ?	58	48.33	46	38.3
4.	Is it necessary to use lime in fish culture ?	114	95	108	90
5.	How do you correct acidic condition of fish culture pond/tank ?	98	81.67	75	62.5
6.	Do you know the recommended dosage of lime used in general ?	93	77.5	65	54.2
7.	Should we have to manure the fish culture ponds ?	114	95	109	90.8
8.	What are the advantages of manuring fish culture pond?	82	68.33	69	57.5
9.	Name some common organic manures used in fish culture	115	95.83	104	86.7
10.	Do you know the rate of application of cow dung (including initial dose and subsequent monthly doses) ?	96	80	75	62.5
11.	How many days before of stocking of fish seed manure should be applied ?	32	26.67	35	29.2
12.	Is it necessary to use inorganic fertilizers in addition to organic manures in fish culture ?	65	54.17	72	60
13.	What are the advantages of using inorganic fertilizer ?	44	36.67	32	26.7
14.	Is it necessary to eradicate excess aquatic weeds ?	111	92.5	94	78.3
15.	Name some aquatic weeds	115	95.83	91	75.8
16.	Do you feel predatory and weed fishes are desirable in fish culture pond ?	113	94.17	97	80.8
17.	Mention any two predatory and two weed fishes known to you.	109	90.83	91	75.8

Contd...

Table 23–*Contd...*

Sl.No.	Practices	Response Category (Yes)			
		North 24 Parganas		Dakshin Dinajpur	
		Freque-ncy	Percen-tage	Freque-ncy	Percen-tage
18	What in the manual method of eradication/control of predatory and weed fishes ?	65	54.17	45	37.5
19.	Name any piscicide used in fish culture	57	47.5	35	29.2
20.	Do you know the recommended dosage of mohua oil cake or bleaching powder ?	51	42.5	32	26.7
21.	Name three Indian major carps	114	95	102	85
22.	Name three exotic carps	75	62.5	61	50.8
23.	Which is the fastest growing major carp and exotic carp ?	71	59.17	78	65
24.	Catla and silver carp are surface feeders.	65	54.17	42	35
	Rohu is a column feeder				
	Mrigal feeds on bottom vegetation				
	Common carp is omnivorous				
25.	Which fish grows well in weed infested ponds?	58	48.33	93	77.5
26.	What is recommended rate stocking for irrigation tanks when CFC is practiced ?	86	71.67	20	16.7
27.	Do you know the recommended species combination for composite fish culture	93	77.5	72	60
	3 SSP – 400 C : 300 R : 300 M (or 300 CC)				
	4 SSP – 300 C : 250 R : 150 M : 300 CC				
	6 SSP – 150 C : 250 R : 100 M : 200 SC : 100 GC : 200 CC				
28.	What is the ideal size of fish seed for stocking ?	67	55.83	70	58.3
29.	Whether supplementary feeding is necessary in CFC?	104	86.67	92	76.7
30.	Name the commonly used supplementary feeds.	111	92.5	93	77.5
31.	What is the best method of feeding ?	81	67.5	65	54.2
32.	Generally supplementary feeding in provided at	46	38.33	21	17.5
33.	After stocking, once in how many days should manuring be done ?	87	72.5	96	80
34.	Do you know the recommended manuring schedule to be practiced after stocking ?	102	85	75	62.5
35.	Do you known the indicators of oxygen depletion in fish pond ?	105	87.5	97	80.8
36.	Is it necessary to stop manuring and feeding when pond water turns greenish ?	87	72.5	74	61.7
37.	Name any fish disease that occurs in fish culture ponds	104	86.67	65	54.2
38.	How do you control disease outbreaks ?	78	65	62	51.7
39.	Is it necessary to check the growth after stocking ?	47	39.17	57	47.5
40.	In general, after how many months of stocking should the fish crop be harvested ?	56	46.67	42	35
41.	What should be the optimum size of harvesting ?	67	55.83	45	37.5

Further, the level of correct knowledge of fish farmers in general, in respect of different items of fish farming varies from item to item, though ranging from slightly above 50 per cent to nearly sixty six per cent. For example, the level in question varies as follows for certain items: remedies for acidic condition of pond water (62.5 per cent), correct dose of lime (54.2 per cent), advantages of manuring (57.5 per cent), dose of organic manure to be applied (62.5 per cent), need for use of inorganic fertilizer (60 per cent), name of the exotic carps (50.8 per cent), fastest growing major carp and exotic carps (65 per cent), species combination in composite fish culture (60 per cent), ideal size of fish seed for stocking (58.3 per cent), recommended manuring schedule to be practised after stocking (62.5 per cent), feeding method (54.2 per cent), water quality management (necessity to stop manuring and feeding) (61.7 per cent), names of the fish disease (54.2 per cent) and disease control method (51.7 per cent).

However, it is clear from table that the majority of the fish farmers lack correct information of such important matters as the on knowledge items pertaining to the minimum depth of water, nutrients required for production of fish food organism, the advantages of using inorganic fertilizers, the names of piscicides, application of mohua oil cake, the method of eradication of predatory and weed fishes, the optimum size of harvesting, harvesting period, the necessity of checking growth.

It is clear from the table that only a few farmers have knowledge about correct feeding rate of supplementary feed (17.5 per cent) and recommended stocking density of different fish species (16.7 per cent).

Thus, it may be concluded that although the fish farmers of the two districts in question are aware of routine and general practices of fish farming, the level of their knowledge in respect issues and matters connected with scientific fish culture may be said to be far from satisfactory. As such their productivity has adversely, been affected by this lack of knowledge in crucial areas, which in turn has tended to stand in the way of profitable pisciculture. The reason may be low education, lack of adequate scientific curriculum in the training programme, poor communication characteristictics etc. Similar findings were reported by Praveena (1993) with respect to the rate of application of manures, fertilizers and detection of diseases. Mahandra Kumar (1996) reported that 38 to 58 per cent of the respondents had correct knowledge on pond management and pond preparation issues which involved scientific study aspects.

Situational Variables

21) Size of Water Body

North 24 Parganas

With respect to the size of water body, the data reveal that most of the fish farmers *i.e.* 50 (41.7 per cent) use a medium sized pond, followed by 45 (37.5 per cent) of fish farmers using low sized ponds while the rest use large sized ponds.

Dakshin Dinajpur

With respect to the size of water body, the data reveals that a large section of

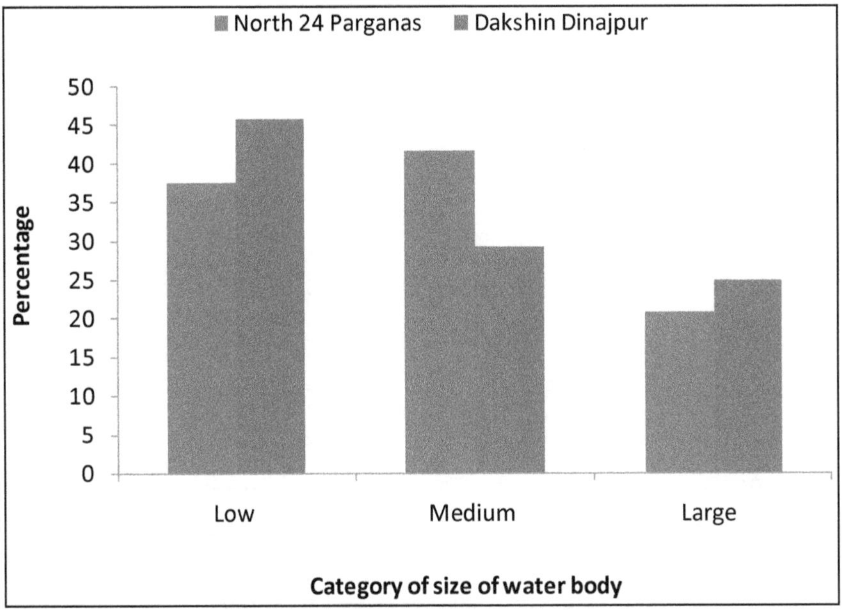

Figure 18: Distribution of Respondents According to Size of Water Body.

fish farmers, *i.e.* 55 (45.8 per cent) belong to the category of low sized pond users followed by 35 (29.2 per cent) of fish farmers belonging to medium sized pond users category while the rest belong to large sized pond users category.

Table 24: Distribution of Respondents According to Size of Water Body

Category	*North 24 Parganas*		*Dakshin Dinajpur*	
	Frequency	*Percentage*	*Frequency*	*Percentage*
Low sized pond	45	37.5	55	45.8
Medium sized pond	50	41.7	35	29.2
Large sized pond	25	20.8	30	25.0

The results indicate that the preponderant size of the fish ponds used is medium in North 24 Parganas while in Dakshin Dinajpur the majority of fish ponds used are low sized. Ponnappan (1982) reported that most of the fish farmers possessed medium sized pond. In another study about 50 per cent of fish farmers in Madhya Pradesh were found operating small farms upto one hectare size (Singh and Nandi, 2001).

22) Duration of Water Availability

North 24 Parganas

The results indicate that the majority of the ponds 91.7 per cent are of the category of long seasonal ponds followed by 8.3 per cent short duration seasonal ponds

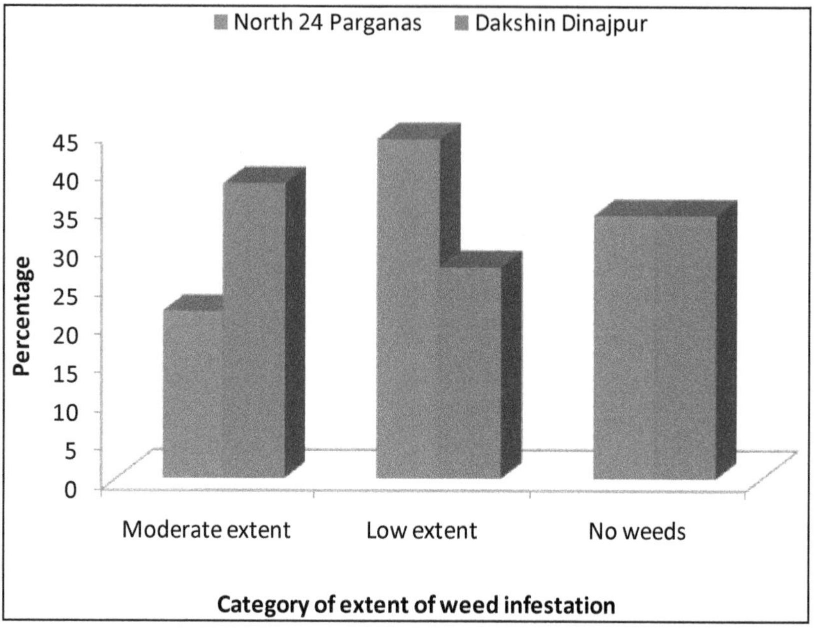

Figure 19: Distribution of Respondents According to Extent of Weed Infestation.

Dakshin Dinajpur

The results show that the majority of the ponds 78.3 per cent are of the category of long seasonal ponds followed by 21.7 per cent short seasonal ponds.

Table 25: Distribution of Respondents According to Duration of Water Availability

Category	*North 24 Parganas*		*Dakshin Dinajpur*	
	Frequency	*Percentage*	*Frequency*	*Percentage*
Short seasonal tanks (6-9 months)	10	8.3	26	21.7
Long seasonal tanks (9-12months)	110	91.7	94	78.3

Most of the ponds of the fish farmers of the two districts belong to long seasonal shower. This is corroborated by the results of the study by Singh and Nande (2001).

23) Source of Water

North 24 Parganas

The data reveal that the majority of the ponds, *i.e.*, 116 (96.7 per cent) are dependent exclusively on rain water. Ponds in receipts both canal and rain water constitute a meagre percentage.

Dakshin Dinajpur

The data reveal that the majority of the ponds, *i.e.*, 112 (93.3 per cent) are dependent exclusively on rain water. A negligible portion of ponds receives both canal and rain water.

Table 26: Distribution of Respondents According to Source of Water

Category	*North 24 Parganas*		*Dakshin Dinajpur*	
	Frequency	*Percentage*	*Frequency*	*Percentage*
Rain water	116	96.7	112	93.3
Canal	0	0	0	0
Both rain and canal	4	3.3	8	6.7

Big and small ponds are built basically to store rain water for domestic and irrigation purposes of the community. It is found that availability of water in the majority of ponds of the two districts is dependent on monsoon rains and ponds receiving water from other sources like canals, etc., are relatively less numerous (7 per cent). This implies that monsoon rains determine the success or failure of fish culture activities to a great extent in the areas under study. Hence it is necessary that extension activities should be focused on rain water conservation measures for fish farming. The present findings are in conformity with those of Singh and Nande (2001).

24) Extent of Weed Infestation

North 24 Parganas

It is observed from the table that the majority of the ponds are infested with aquatic weeds in varying degrees. A greater number representing 53.0 per cent of ponds are found infested with a lower quantity of weeds, followed by infestation free ponds (34.2 per cent) and ponds affected by moderate infestation (21.7 per cent).

Dakshin Dinajpur

It is observed from the table that majority of the ponds are infested with aquatic weeds in varying degrees. The largest section of ponds 38.3 per cent are found infested to a medium extent, while 34.2 per cent of the ponds are found to be free from weed infestation altogether.

Table 27: Distribution of Respondents According to Extent of Weed Infestation

Category	*North 24 Parganas*		*Dakshin Dinajpur*	
	Frequency	*Percentage*	*Frequency*	*Percentage*
Moderate extent	26	21.7	46	38.3
Low extent	53	44.1	33	27.5
No weeds	41	34.2	41	34.2

Aquatic weeds in a negligible quantity are desirable in fish culture ponds. It becomes a menace when they are present in large quantities. The weeds create management problems to a large extent. Hence, the extent to which aquatic weeds should be present in a water body is an important issue to be decided in management and should carefully be sorted out for better management and successful adoption of recommended practices. The results reveal that the majority of ponds are infested with weed within a manageable limit while one third of the ponds are free from aquatic weeds. Hence it can be concluded that most of the water bodies of the two districts are weed infested in variable degrees.

Dependent Variables

Attitude of Fish Farmers towards Scientific Fish Culture

North 24 Parganas

Almost an equal percentage of fish farmers belong to either medium favourable attitude category (40.8 per cent) or more favourable attitude category (35 per cent), as to their response towards scientific fish culture practices, which in turn indicates a positive attitude amongst the greater number of fish farmers. However, 22.5 per cent of the respondents (25.8 per cent) show a less favourable attitude. The mean attitude scores of the less favourable, favourable and more favourable categories are 61.3, 66.5 and 78.2 respectively.

Dakshin Dinajpur

Almost an equal percentage of fish farmers belong either to favourable attitude category (37.5 per cent) or more favourable attitude category (33.3 per cent), as to their response towards scientific fish culture practices indicating positive attitude amongst the greater portion of fish farmers. However, 29.2 per cent of the respondents (25.8 per cent) show a less favourable attitude. The mean attitude scores of the less favourable, favourable and more favourable categories are 54.5, 65.1 and 72.6 respectively.

Table 28 : Distribution of Respondents According to Level of Favourable Attitude towards Scientific Fish Culture

Attitude Categories	*North 24 Parganas*			*Dakshin Dinajpur*		
	Frequency	*Percentage*	*Mean Attitude Scores*	*Frequency*	*Percentage*	*Mean Attitude Scores*
Less favourable	27	22.5	61.3	35	29.2	54.5
Medium favourable	49	40.8	66.5	45	37.5	65.1
More favourable	44	36.7	78.2	40	33.3	72.6

The data reveal that the majority of the respondents of the two districts have medium to more favourable attitude towards scientific fish culture practices. The plausible explanation for favourable attitude of the respondents towards scientific fish culture is that the age group of the majority of them varies from young to middle aged which is conducive to their being responsive to new ideas on fish culture practices. Acquisition of skills has given them better confidence, which in turn has led to a favourable attitude. These findings are in agreement with those of Ponnappan (1982), Balasubramaniam (1988), Haque (1989), Mahandra Kumar (1996) and Talukder (2000). Further, another favourable factor of training has done

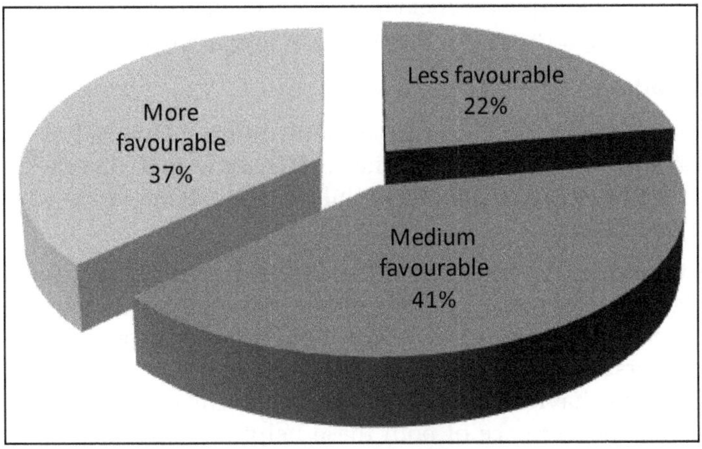

North 24 Parganas

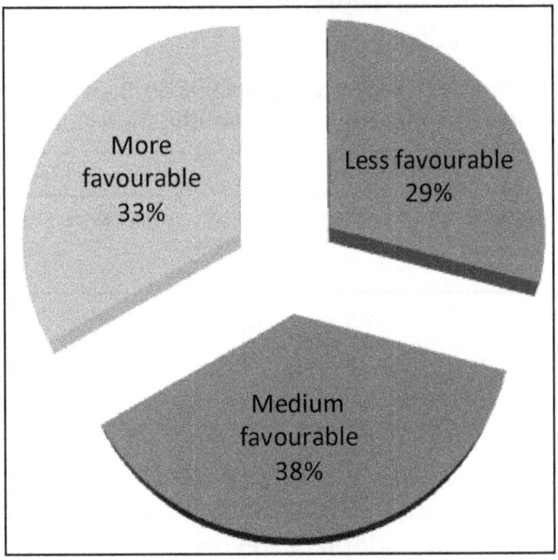

Dakshin Dinajpur

Figure 20: Distribution of Respondents Based on their Level of Attitude in North 24 Parganas and Dakshin Dinajpur.

a lot to brining about a suitable change in the attitude of the trainees towards prawn farming, as also viewed by Basava Kumar and Yaligar (2000).

Adoption Behaviour of Fish Farmers towards Scientific Fish Culture

a) Overall Adoption

North 24 Parganas

The average adoption score achieved by the low, medium and high categories is 62.2, 79.8 and 88.6 respectively. The results show that as large a portion as representing almost 82.5 per cent belongs to the category of medium to high adopters while the remaining farmers belong to the low adopter category.

Dakshin Dinajpur

The mean of the adoption scores of low, medium and high adoption categories is 55.2, 71.3, and 79.5 respectively. The results show that almost 70 per cent of farmers are medium to low adopters while the remaining farmers belong to the high adoption category.

In both the districts the status of the adoption behaviour of the fish farmers towards scientific fish culture is on the whole encouraging, though the position is somewhat more satisfactory so far as North 24 Parganas is concerned. The attitude of the fish farmers in this regard is also correspondingly favourable to improvements in their adoption behaviour-a sort of symbiotic interrelationship being noticeable. Taken together the presence of both these helpful factors is very important in promoting a general climate on economic well being and entrepreneurial enthusiasm as based on the progress of the exploitation of the potentials of a kind of natural resources in the right direction.

Table 29: Distribution of Respondents Based on their Level of Adoption of Fish Farmers towards Scientific Fish Culture

Adoption Categories	*North 24 Parganas*			*Dakshin Dinajpur*		
	Frequency	*Percentage*	*Mean Adoption Scores*	*Frequency*	*Percentage*	*Mean Adoption Scores*
Low	21	17.5	62.2	40	33.3	55.2
Medium	76	63.3	79.8	44	36.7	71.3
High	23	19.2	88.6	36	30.0	79.5

The results indicate that the majority of the fish farmers belong to medium adopter category (63.3 per cent) in North 24 Parganas whereas in Dakshin Dinajpur they belong to medium to low adopter category. Meeran (1983) reported medium to high level of adoption. Mahandra Kumar (1996) concluded that the majority of fish farmers adopted composite fish culture technologies only to a medium extent. The results indicate that the majority of the fish farmers of North 24 Parganas have adopted the scientific fish culture technology to a medium extent while in the case of Dakshin Dinajpur the adoption level ranges from medium to low, which might be due to the reason that most of the farmers of this area has precise knowledge about many of the simple and basic practices of scientific fish culture alone. Although this is an encouraging trend, efforts are still required to promote large scale adoption of this technology.

The present findings are more or less in conformity with those reported by Subhas Chandra (1986), Krishnaiah (1989) and Bhaumik *et al.* (1992). However, they differ from the the findings of the study made by Sen and Das (1986), Bose (1989), Wayal (1990), Magle *et al.* (1991), Praveena (1993), Mahandra Kumar (1996), Srinath (1996), Bhaumik and Saha (1998), Panigrahi and Dana (2002).

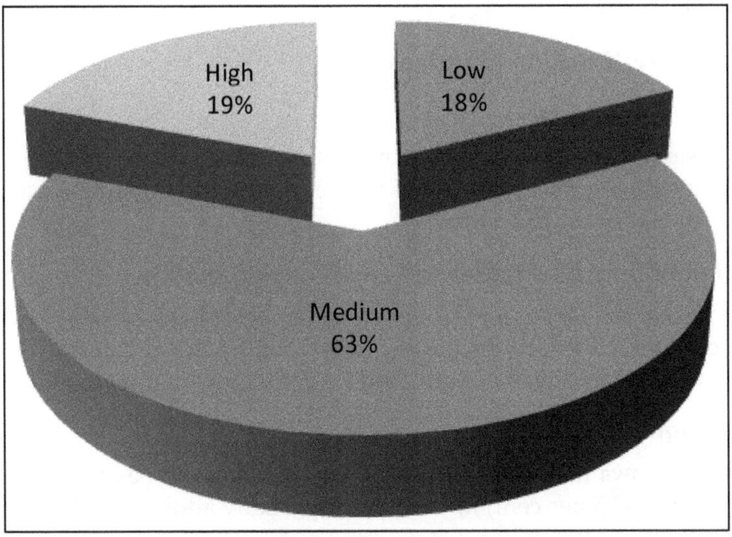

North 24 Parganas

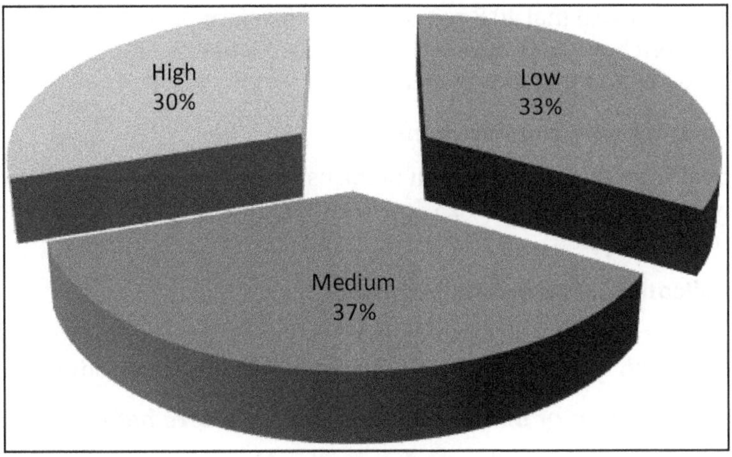

Dakshin Dinajpur

Figure 21: Distribution of Respondents Based on their Level of Adoption in North 24 Parganas and Dakshin Dinajpur.

b) Extent of Adoption of Specific Recommended Scientific Fish Culture Practices by Fish Farmers

In the interest of a better understanding of the adoption behaviour, the recommended practices of scientific fish culture are categorized into 3 groups, namely, pre stocking practices, stocking practices and post stocking practices. Accordingly, data collected under these heads are presented below.

North 24 Parganas

Adoption of Pre-Stocking Practices

Testing of soil and water

It is observed that 36.7 per cent of the fish farmers have partially adopt the practice followed by non adopters (35 per cent), and full adoption accounting for only 28.3 per cent of the fish farmers.

Application of lime for improving soil and water quality

Most of the fish farmers (45.8 per cent) have adopted partially this practice, 37.5 per cent of the fish farmers having fully adopted this practice, while 16.7 per cent of the fish farmers have not adopted lime application practice.

Control of aquatic weeds

Table-30 shows that 42.5 per cent of the fish farmers have not adopted this practice. About (37.5 per cent) of them have partially adopt and 20 per cent alone fully adopting this practice.

Control of predators and undesirable fish species

The results reveal that 40.8 per cent of the fish farmers have partially adopted and 20.8 per cent have fully adopted this practice whereas 38.4 per cent of the fish farmers have not adopted this practice.

Initial manuring with organic manures

It is clear from the result that half of the fish farmers (50 per cent) have partially adopt this practice, followed by 23.3 per cent of full adoption section and 26.7 per cent case of nil adoption.

Initial application of inorganic fertilizers

The majority of the fish farmers (52.5 per cent) have partially adopted this practice. However, 32.5 per cent of fish farmers fall under the non adoption category.

Period of application of initial manures and fertilizers before stocking :

These findings relate to the degree of adherence to the prescribed periodicity of application of manures and fertilizers. The results reveal that 60 per cent of fish farmers fall under full adoption category followed by 27.3 per cent belonging to the partial adoption category, while 11.7 per cent of the fish farmers are found to be reluctant to follow the time schedule in applying manures and fertilizers. The above results indicate favourable acceptance of this practice by the majority of fish farmers.

Stocking Practices

Stocking of recommended number of different fish species

The data indicate full adoption by the majority (47.5 per cent) of the fish farmers have only, followed by partial adoption by 45.8 per cent of them.

Density of stocking fish seed (number/ha)

The majority of fish farmers (46.7 per cent) are full adopters of this practice. However, 43.3 per cent of them belong to the category of partial adoption of the practices.

Ratio of stocking of selected species

A major portion of fish farmers (47.5 per cent) falls under partial adoption category followed by full adoption by 36.7 per cent and non adoption by 15.8 per cent of fish farmers.

Size of fish seed stocked

The majority (61.7 per cent) of the fish farmers have fully adopted the correct size of stocking fish seed followed by partial adoption by 28.3 per cent. However, a few farmers (10 per cent) are non adopters.

Post Stocking Practices

Use of supplementary feeds in prescribed proportion

It is clear from the result that 45 per cent of the fish farmers have fully adopt this practice, followed by partial adoption (43.3 per cent). However, 11.7 per cent of the fish farmers do not apply the feeds during fish culture practices.

Rate of feeding

This practice concerns to those farmers who adopt the use of supplementary feeds in prescribed proportion. Some 13.2 per cent of the respondents do not adopt this practice, while 68.4 per cent of them adopt this practice partially. However, a small portion of fish farmers (18.4 per cent) are adopted the practices in full *i.e.* full adopters.

Feeding method

This practice relates to those farmers who adopt the use of supplementary feeds in the prescribed proportion. With regard to feeding method, 10.5 per cent of the respondents do not adopt this practice, while 23.7 per cent of them fully adopt the practice. However, majority of the farmers (65.8 per cent) are found to be partial adopters.

Post stocking manuring with organic manures

Most of the fish farmers (38.3 per cent) fully adopt post stocking manuring with organic manures, while 34.2 per cent of fish farmers adopt this practice partially and the rest (27.5 per cent) do not apply any organic manure, during the post stocking period.

Post stocking manuring with inorganic fertilizers

It is found that the fish farmers are almost equally divided between the categories of partial adoption or non adoption, while 29.2 per cent belong to the category of full adoption.

Periodic netting to check growth and health of fish

It is observed that nearly half of the fish farmers (52.5 per cent) partially adopt the practice of periodic netting, followed by full adopters (27.5 per cent), while non adoption accounted for only 20 per cent of the fish farmers.

Taking into consideration the facts and figures furnished above that there is quite a high percentage of partial adopters amongst the fish farmers with respect to scientific fish culture practices like stocking of recommended number of different fish species, density of stocking of fish seed of different fish species, ratio of stocking of selected species, use of supplementary feeds in prescribed proportion rate of feeding, feeding method, initial manuring with organic manures, application of inorganic fertilizer and periodic netting.

Table 30: Adoption of Recommended Scientific Fish Culture Practices in North 24 Parganas

Sl.No.	Recommended Practices	Fish Farmers					
		Full Adoption		Partial Adoption		Non adoption	
		No.	%	No.	%	No.	%
A.	**Pre stocking practices**						
1.	Testing of soil and water	34	28.3	44	36.7	42	35
2.	Application of lime for improving soil and water quality	45	37.5	55	45.8	20	16.7
3	Control of aquatic weeds	24	20	45	37.5	51	42.5
4.	Control of predators and undesirable fish fauna	25	20.8	49	40.8	46	38.4
5.	Initial manuring with organic manures (RCD or poultry).	28	23.3	60	50	32	26.7
6.	Application of inorganic fertilizer (urea and super phosphate)	18	15	63	52.5	39	32.5
7.	Period of application of initial manures and fertilizer before stocking (n=110)	66	60	30	27.3	14	11.7
8.	Stocking of recommended number of different fish species	57	47.5	55	45.8	8	6.7
9.	Density of stocking of fish seed (no./ha)	56	46.7	52	43.3	12	10
10.	Ratio of stocking of selected species	44	36.7	57	47.5	19	15.8
11.	Size of the fish seed stocked	74	61.7	34	28.3	12	10
12.	Use of supplementary feeds in prescribed proportion	54	45	52	43.3	14	11.7
13.	Rate of feeding (n=76)	14	18.4	52	68.4	10	13.2
14.	Feeding method (n=76)	18	23.7	50	65.8	8	10.5
15.	Post stocking manuring with organic manures	46	38.3	41	34.2	33	27.5
16.	Post stocking fertilization with inorganic fertilizers	35	29.2	43	35.8	42	35
17.	Periodic netting to check the growth and health of fishes	33	27.5	63	52.5	24	20

Figures in parenthesis indicate number of adopters

The phenomenon of full adoption amongst the fish farmers is observed only in respect of a few practices such as periodic application of manures and fertilizers, size of fish seed stocked. Further, it is also observed that the majority of the fish farmers do not at all adopt the practices such as testing of soil and water, control of aquatic weeds, control of predators and undesirable fish fauna and post stocking fertilization with inorganic fertilizers.

Dakshin Dinajpur

Adoption of Pre-stocking Practices

Testing of soil and water

A very small portion *i.e.* 3.3 per cent of the fish farmers either partially or fully adopt this practice.

Application of lime for improving soil and water quality :

A great majority (65 per cent) of the fish farmers do not adopt this practice at all while nearly 1/3 of them (30.8 per cent) adopt partially and a very small portion of them (4.2 per cent) fully adopt lime application practice.

Control of aquatic weeds

Table 31 shows that nearly three quarters of the fish farmers (70 per cent) adopt this practice partially (in varying degrees) while a small number (5.8 per cent) of fish farmers adopt it to a full extent and nearly a quarter of fish farmers (24.2 per cent) do not adopt this practice at all. The fact of partial adoption in such a large scale may be attributed to various practical difficulties in the control of aquatic weeds in water bodies.

Control of predators and undesirable fish species

The results reveal that the majority of fish farmers adopt this practice. Those who partially adopt this practice accounts for about (80.8 per cent) and those who fully adopt it constitute 1.7 per cent, 17.5 per cent of them do not adopt this practice at all.

Initial manuring with organic manures

As regards to the application of organic manures, the majority of the fish farmers (56.7 per cent) do not adopt the practice altogether while 40 per cent of them adopt it partially and a very small fraction (3.3 per cent) adopt this practice fully.

Initial application of inorganic fertilizers

The majority of the fish farmers (55.8 per cent) do not adopt this practice, while the rest of them 44.2 per cent adopt it partially, being given to applying less or more than the recommended doses of fertilizer.

Period of application of initial manures and fertilizers before stocking

These findings relate to the degree of adherence to the prescribed periodicity of application of manures and fertilizers. The results reveal that 53 per cent of fish farmers fall under full adoption category followed by those falling under the category of partial adoption (24.5 per cent), while 22.4 per cent of the fish farmers are reluctant to follow the time schedule while applying manures and fertilizers.

The above results indicate favourable acceptance of this practice by the majority of the fish farmers.

Stocking Practices

Stocking of recommended number of different fish species

The data indicate that a great majority (81.7 per cent) of fish farmers only partially follow this important practice of stocking of recommended number of different fish species, followed by full adoption by the remaining fish farmers, who, however, represented a meagre percentage (18.3 per cent) of them.

Density of stocking fish seed (number/ha)

The majority of the fish farmers (70 per cent) are partial adopters of this practice. However, the rest of them (30 per cent) belong to the category of full adoption of the practices.

Ratio of stocking of selected species

A high percentage (68.3 per cent) of fish farmers fall under partial adoption category followed by non adoption by 20 per cent and full adoption by 11.7 per cent of fish farmers.

Size of fish seed stocked

The majority of the fish farmers (58.3 per cent) fully adopt correct size while stocking fish seed followed by partial adoption by 28.3 per cent. However, a few farmers (13.4 per cent) are found to be non adopters.

Post Stocking Practices

Use of supplementary feeds in prescribed proportion

It is clear from the assessment that nearly half of the fish farmers (50.8 per cent) partially adopt this practice, followed by full adoption by 6.5 per cent. However, 42.5 per cent of fish farmers do not apply the feeds during the fish culture practices.

Rate of feeding

This practice is adopted by those farmers who make use of supplementary feeds in the prescribed proportion. 7 per cent of the respondents do not adopt this practice. 90 per cent of fish farmers adopt this practice partially. However, a very small portion of fish farmers (3 per cent) are found to have adopted the practice in full.

Feeding method

This practice is current among those farmers who adopt the use of supplementary feeds in the prescribed proportion. With regard to feeding methods, 37.6 per cent of the respondents do not adopt this practice, while 3 per cent of them partially adopt the practices. However, the majority of the farmers (59.4 per cent) are full adopters.

Table 31: Adoption of Recommended Scientific Fish Culture Practices in Dakshin Dinajpur

Sl.No.	*Recommended Practices*	*Fish Farmers*					
		Full Adoption		*Partial Adoption*		*Non adoption*	
		No.	*%*	*No.*	*%*	*No.*	*%*
A.	**Pre stocking practices**						
1.	Testing of soil and water	4	3.3	4	3.3	112	93.4
2.	Application of lime for improving soil and water quality	5	4.2	37	30.8	78	65
3	Control of aquatic weeds	7	5.8	84	70	29	24.2
4.	Control of predators and undesirable fish fauna	2	1.7	97	80.8	21	17.5
5.	Initial manuring with organic manures (RCD or poultry).	4	3.3	48	40	68	56.7
6.	Application of inorganic fertilizer (urea and super phosphate)	0	0	53	44.2	67	55.8
7.	Period of application of initial manures and fertilizer before stocking (n=98)	52	53.1	24	24.5	22	22.4
B.	**Stocking practices**						
8.	Stocking of recommended number of different fish species	22	18.3	98	81.7	0	0
9.	Density of stocking of fish seed (no./ha)	36	30	84	70	0	0
10.	Ratio of stocking of selected species	14	11.7	82	68.3	24	20
11.	Size of the fish seed stocked	70	58.3	34	28.3	16	13.4
C.	**Post stocking practices**						
12.	Use of supplementary feeds in prescribed proportion	8	6.5	61	50.8	51	42.5
13.	Rate of feeding (n=69)	2	3	62	90	5	7
14.	Feeding method (n=69)	41	59.4	2	3	26	37.6
15.	Post stocking manuring with organic manures	6	5	91	75.8	23	19.2
16.	Post stocking fertilization with inorganic fertilizers	0	0	36	30	84	70
17.	Periodic netting to check the growth and health of fishes	21	17.5	65	54.2	34	28.3

Figures in parentheses indicate number of adopters

Post stocking manuring with organic manures

The great majority of the fish farmers *i.e.,* 75.08 per cent partially adopt the post stocking manuring with organic manures, while 19.2 per cent of fish farmers adopt this practice fully and the rest (5 per cent) are found to be applying no organic manure during the post stocking period.

Post stocking manuring with inorganic fertilizers

It is found that a majority (70 per cent) of the fish farmers do not use inorganic fertilizer and only 30 per cent of them adopt this practice partially.

Periodic netting to check growth and health of fish

It is observed that nearly half of the fish farmers (54.17 per cent) partially adopt the practice of periodic netting followed by non adopters (28.33 per cent) while full adoption accounted for only 17.5 per cent of the fish farmers.

The above pieces of information lead to the conclusion that quite a large number of farmers partially adopt scientific fish culture practices, such as control of aquatic weeds; control of predators and undesirable fish species, stocking of recommended number of different fish species, stocking density, ratio of stocking, use of supplement feeds and post stocking manuring with organic manures and periodic netting.

The trend of full adoption among the fish farmers is observed only in respect a few practices such as periodic application of manures and fertilizers, size of fish seed stocked and feeding method. Further it is also observed that the majority of fish farmers did not adopt the practices such as testing of soil and water, application of lime for improving soil and water quality, initial manuring with organic manure, and the use of inorganic fertilizer before stocking and after stocking.

The results indicate wider adoption of application of lime for improving soil and water quality by the majority of the fish farmers in North 24 Parganas as against the lack of such application by the majority of the fish farmers in Dakshin Dinajpur. These findings of Dakshin Dinajpur are in agreement with the findings of Mahandra Kumar (1996) and Mishra and Mishra (1999).

The plausible reason for lack of adoption of soil and water testing practices in Dakshin Dinajpur is lack of knowledge coupled with non-availability of the facilities of soil and water testing for adoption. Balasubramaniam (1988) reports non-adoption of this practice by the farmers to an extent of 65 per cent.

It is observed that the practice of control of aquatic weeds is adopted to a lesser extent by the majority of the fish farmers of North 24 Parganas as compared to Dakshin Dinajpur. The respondents of Dakshin Dinajpur adopted this practice partially as to facilitate easy harvesting. Balasubramaniam (1988) reported partial adoption of this practice. However, Krishnaiah (1989) reports full adoption of the practice.

The practice of control of predators and undesirable fish fauna is partially adopted in the form of employing repeated netting by the most of the fish farmers. The present findings contradict the findings reported by Bhaumik *et al.* (1992) but are in conformity with those of Santosh and Sudhir (1992)

The data reveal that the majority of the respondents of Dakshin Dinajpur are non-adopters, and partial adoption is noticed in the case of 40 per cent of the farmers where only a very low dose of organic manures is used. As against this in North 24 Parganas the majority of the fish farmers are found to adopt this practice a partially. The findings are in conformity with the findings reported by Meeran (1983). The probable reason for use of low dosage of organic manure may be the non availability of sufficient quantities of organic manure and their primary preference for using the limited supply of this costly resource in agricultural lands for crop cultivation rather than fish farming.

Data indicate that almost none of the fish farmers use inorganic fertilizers like urea and super phosphate for seed preparation. Similar observations occur in an earlier study by Meeran (1983).

The correct timing of application of initial manures and fertilizers before stocking operation is maintained by those farmers who apply organic manures and fertilizers (n=98). The adherence to the right period of application of manures enhances the survivability of the stocked fish seed. The data reveal that nearly 53 per cent of the fish farmers who apply organic manures and fertilizers adopt the practice fully, followed by partial adoption by 24.5 per cent of the fish farmers.

The results indicate that majority of the fish farmers partially adopt stocking of recommended number of different fish species wherein stocking of different fish species range from one to three recommended species. Meeran (1983) reports adoption of stocking of culturable species to an extent of more than 75 per cent level while poor adoption of the practice is reported by Bhaumik *et al.* (1992) and Santhosh and Sudhir (1992). However, the findings reported by Krishnaiah (1989) and Mahandra Kumar (1996) do not support the results.

Regarding stocking density of fingerlings, the majority of the fish farmers (81.7 per cent) are partial adopters. In West Bengal, only 10 per cent of the fish farmers practise appropriate stocking rate (Mishra, 1987). The results are in line with the findings reported by Subhash Chandra (1986) and Santosh and Sudhir (1992). Meeran (1983) observes adoption of proper stocking density of fingerlings by nearly $1/3^{rd}$ of the respondents.

With respect to the ratio of stocking, it is found that the majority of farmers are partial adopters. It shows that the majority of them deviate from the recommended ratio. However, a very small section of farmers belong to both full and non-adoption categories. It is noticed that the failure to get supply of the desired fish species timely is a major problem faced by the farmers, which is responsible for non adoption of proper stocking ratio. Similar findings are reported by Bhaumik *et al.* (1992) whereas Mahandra Kumar (1996) reports full adoption by the majority of farmers.

The results reveal that the majority of the fish farmers are aware of the correct quantum of fish seed to be stocked. Nearly 58.3 per cent of fish farmers are full adopters followed by partial adopters (28.3 per cent). Hence, it may be concluded that this practice is on the whole favourably accepted by the fish farmers. However, the findings of study of Mahandra Kumar (1996) do not support the present findings. Santosh and Sudhir (1992) reports stocking of fry of cultivable species in Orissa instead of fingerlings.

It is observed from the data that the recommended feeds in prescribed ratio (rice bran: mustard oil cake 1:1) are used by a small fraction (6.5 per cent) of fish farmers alone whereas partial adoption is preferred by the majority of the fish farmers. It may be inferred that fish farmers use locally available and cheap ingredients in different ratio so as to economize the cost of feed. Work of Balasubramaniam (1988) supports the present findings.

It is observed from the data that 3.1 per cent of the farmers adopt the practice relating to the rate of feeding. The majority of the farmers partially adopt this practice. Similar findings are reported by Subhas Chandra (1986), Santosh and Sudhir (1992) and Mishra and Mishra (1999).

The results reveal that the recommended feeding methods are practised by more than half of fish farmers (50.9 per cent), while the rest deviate from the recommended method.

The results reveal that the majority of the farmers partially adopt post stocking manuring with organic manures. Lack of knowledge and villagers' objection to manuring the ponds are the predominant factors responsible for partial adoption. Occasional manuring and low application of manures are reported by Santosh and Sudhir (1992) and Mishra and Mishra (1999).

It is observed from the data that the practice of using inorganic fertilizers during the post stocking management is rare. Most of the fish farmers are are non-adopters and 30 per cent of them follow partial adoption (in very low doses).The present findings are in line with the findings reported by Meeran (1983), Santosh and Sudhir (1992) and Mishra and Mishra (1999).

The findings reveal that the majority of the respondents adopt the practice of resorting to periodic netting in order to check the growth and health of the fish, either partially or fully. However, those who adopt partially are predominant in number, which implies that fish farmers are giver to occasionally checking the growth and health of fish by netting operations. Poor awareness coupled with lack of necessary equipments possibly account for the partial adoption and non-adoption. Similar findings are reported by Meeran (1983).

PART-II

Relational Analysis

A. Correlation Coefficient

1. Relationship between Socio-personal, Socio-economic, Communication, Psychological and Situational Variables of Fish Farmers and their Attitude towards Scientific Fish Culture Practices

In order to measure the relationships that exist between the attitude of fish farmers on the one hand and such characteristics of theirs as the socio-personal, socio-economic, communication, psychological and situational factors, the correlation coefficient worked out was tested for its statistical significance and is presented in Table 32.

North 24 Parganas

It is observed from the Table 32 that out of the 24 independent variables studied, 10 variables are found to have a positive and significant correlation with the attitude of the fish farmers towards scientific fish culture practices while the remaining 14 variables are found to have a non significant correlationship.

It is clear from the table that most of the socio-personal and situational variables show a non-significant correlationship in question. To be specific, the variables which show non-significant relationship are age (X_1), caste (X_2), family size (X_3), family type (X_4), fish farming experience (X_6), occupation (X_7), land holding (X_9), possession of fishing equipment (X_{11}), extension agency contact (X_{13}), credit orientation (X_{16}), size of the water body (X_{21}), duration of water availability (X_{22}), source of water (X_{23}) and extent of weed infestation (X_{14}).

Table 32: Relational Analysis of different Independent Variables with the Attitude of Fish Farmers towards Scientific Fish Culture as Dependent Variable in North 24 Parganas (N=120) and Dakshin Dinajpur (N=120) District

Independent Variables	*Correlation Coefficient (r)*	
	Pearson Coefficient	
	North 24 Parganas (N=120)	*Dakshin Dinajpur (N=120)*
Age	–0.147 NS	–0.022 NS
Caste	–0.172 NS	0.0946 NS
Family size	0.134 NS	0.0176 NS
Family type	0.141 NS	–0.213*
Education	0.184*	0.347**
Fish farming experiences	0.097 NS	0.285**
Occupation	0.136 NS	0.106 NS
Annual income	0.350**	0.433**
Land holding	0.146NS	0.241*
Social participation	0.347**	0.428**
Possession of fishing equipment	–0.046NS	0.147 NS
Mass media participation	0.331**	0.436**
Extension agency contact	0.146 NS	0.472**
Cosmopoliteness	0.480**	0.617**
Innovative proneness	0.417**	0.624**
Credit orientation	–0.017 NS	0.152 NS
Value orientation	0.330**	0.546**
Risk orientation	0.237**	0.509**
Economic motivation	0.322**	0.657**
Knowledge	0.621**	0.772**
Size of water body	–0.037 NS	0.300**
Duration of water availability	0.005 NS	–0.016NS
Source of water	0.085NS	0.213*
Extent of weed infestation	0.164 NS	0.275**

NS = Non significant; * = Significant at 0.05 level of probability; ** = Significant at 0.01 level of probability.

The variable education (X_5) has a positive and significant relationship at 5 per cent level, while variables like annual income (X_8), social participation (X_{10}), mass media participation (X_{12}), cosmopoliteness (X_{14}), innovative proneness (X_{15}), value

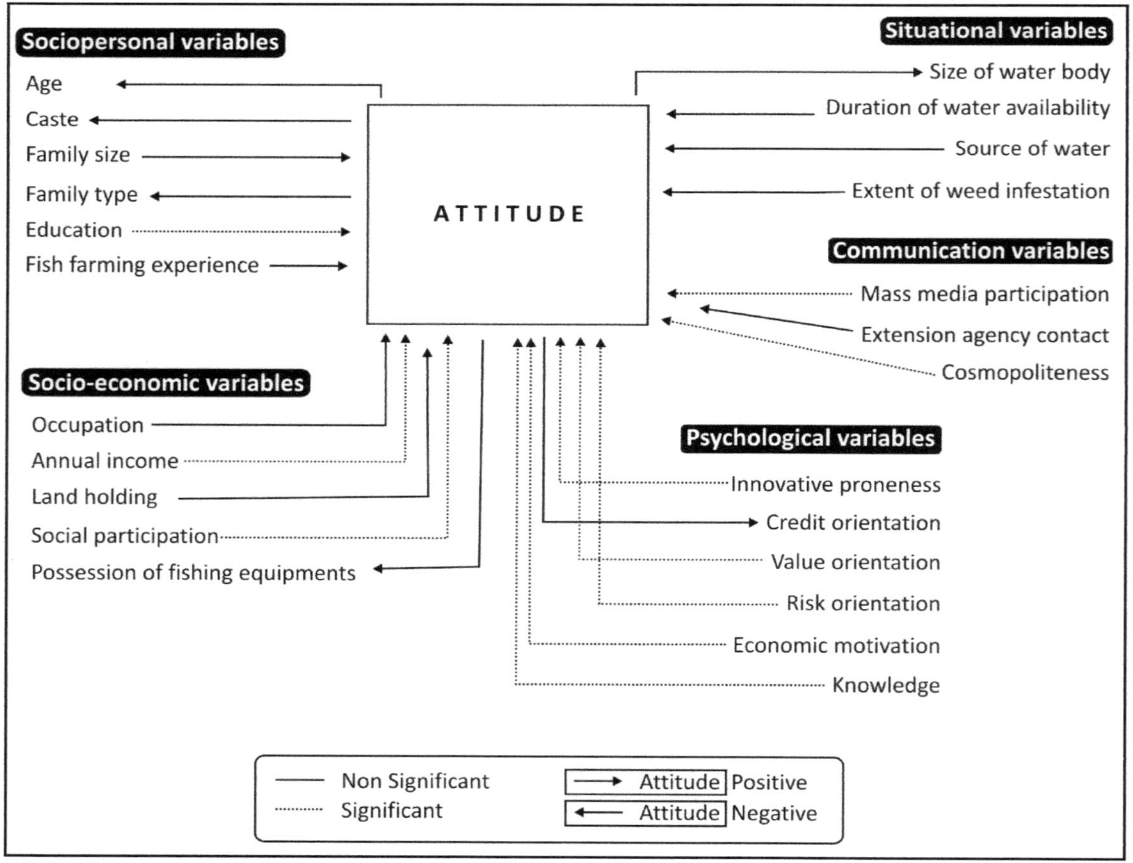

Figure 22: Schematic Diagram Showing the Empirical Relationship between Attitude of Fish Farmers towards Scientific Fish Culture Practices and Independent Variables (North 24 Parganas).

orientation (X_{17}), risk orientation (X_{18}), economic motivation (X_{19}), knowledge about scientific fish culture (X_{20}) have highly positive and significant correlationship with the attitude of the fish farmers at 1 per cent level.

Those variables which have negative correlationship with the attitude of fish farmers include age (X_1), caste (X_2), possession of fishing equipment (X_{11}), credit orientation (X_{16}) and size of the water body (X_{21}).

Dakshin Dinajpur

It is observed from the Table 32 that out of 24 independent variables studied, 16 variables are positively and significantly correlated and only one variable is negatively and significantly correlated with the attitude of fish farmers towards scientific fish culture practices while 7 variables are found to have non-significant correlationship.

It is clear from the Table 32 that most of the personal factors show non-significant correlationship. Age (X_1), caste (X_2), family size (X_3), occupation (X_7), possession of fishing equipment (X_{11}), credit orientation (X_{16}), duration of water availability (X_{22}) have non significant relationship with the attitude of fish farmers towards scientific fish culture practices.

Family type (X_4), land holding (X_9), source of water (X_{23}) have a significant correlationship at 5 per cent level while variables education (X_5), fish farming experience (X_6), annual income (X_8), social participation (X_{10}), mass media participation (X_{12}), extension agency contact (X_{13}), cosmopoliteness (X_{14}), innovative proneness (X_{15}), value orientation (X_{17}), risk orientation (X_{18}), economic motivation (X_{19}), knowledge about scientific fish culture practices (X_{20}) size of water body (X_{21}), extent of weed infestation (X_{24}) show highly positive and significant correlationship with the attitude of fish farmers about scientific fish culture practices at 1 per cent level.

Among the variables which show negative association are age (X_1), family type (X_4), duration of water availability (X_{22}).

Age has negative and non-significant correlationship with the attitude of fish farmers in both the districts. The results are corroborated by the findings of Ponnappan (1982) and Nayametha Krishnan (1988). However, the results of the works of Seehappa and Surendra (1988) contradict the present findings.

It is observed from Table 32 that the variable caste shows a non-significant correlationship with the attitude of fish farmers in the two districts. The results imply that irrespective of caste, the respondents have a similar attitude towards composite fish culture. The information could be utilized by the Development Departments in connection with taking up projects and measures to motivate the resource-rich fish farmers irrespective of caste to take to scientific fish culture.

Family size shows a non-significant relationship with the attitude of the fish farmers towards scientific fish culture in the two districts. The findings are in agreement with the results of the studies by Digrashkar and Wangikar (1990) and Nijagonda (2000) but are contradicted by those of the work by Awasthi *et al.* (2000).

Family type shows a non-significant correlationship with the attitude of the fish farmers in North 24 Parganas whereas in Dakshin Dinajpur the correlationship is negative and significant. The findings reported by Dixit (1988) contradicted the results with regard to North 24 Parganas.

The education levels of the fish farmers show a positive and significant correlationship with their attitude in the two districts. Similar findings are reported by Roy and Pathak (2000).

Experience in fish farming and the attitude of fish farmers exhibit significant and positive correlationship in Dakshin Dinajpur. The farmers possessing some experience about scientific fish culture have a positive attitude. The findings are in agreement with those of Biswas (1990).

It is observed from the table-32 that occupation shows a non-significant relationship with the attitude of fish farmers of the two districts. These findings are in line with those of Navaneetha Krishnan (1988).

Annual income exhibits a positive and significant correlationship with the attitude of fish farmers towards scientific fish culture of the two districts. These findings are in agreement with the observations by Raju Naik *et al.* (1994), Natarajan Muthaiah (1995), Mahandra Kumar (1996), Nisha Arabind (1999) and Awasthi *et al.* (2000).

Land holding is found to be positively and significantly correlated to the attitude of the fish farmers towards scientific fish culture in Dakshin Dinajpur. The findings are in conformity with the findings of Raju Naik *et al.* (1994), Nimbal (1995), Surekha *et al.* (1997) and Nijagonda (2000).

A positive and significant correlationship is found to exist between social participation and attitude of the fish farmers of the two districts towards scientific fish culture. The findings in question are in agreement with the results obtained by Prasad (1995), Mahandra Kumar (1996) and Nijagonda (2000). This might be due to the fact that the respondents with high social participation come from different social levels with diversified experiences and seek to widen the horizon of enterprise and thereby develop a favourable attitude.

The variable 'possession of fishing equipment' has no correlation with the attitude of the fish farmers of the two districts. This implies that the issue of possession of fishing equipments by the respondents does not affect their attitude at all and this variable has thus no role in developing a favourable attitude towards scientific fish culture.

The correlationship between mass media participation and the attitude of fish farmers is found to be positively significant in two districts. The instant findings of the study are in line with the study reports by Natarajan and Muthaiah (1995), Prasad (1995) and Nisha Aravind (1999).

Extension Agency Contact shows a positive and significant correlationship with the attitude of the fish farmers of two districts. A high degree of extension agency contact enables the fish farmers to get more information of a useful nature and build up confidence in the enterprise. The findings of the study are in conformity with those of Ponnappan (1982), Navaneetha Krishnan (1988), Natarajan and Muthaiah (1995), Prasad (1995) and Awasthi *et al.* (2008). But, curiously enough, in North 24 Parganas extension agency contact show a non-significant relationship with the attitude of the fish farmers towards scientific fish culture. This implies that irrespective of extension agency contact the farmers of that district develop their attitude, so this variable has no role in developing a favourable attitude on their part towards fish culture.

Cosmopoliteness has a positive and highly significant correlationship with the attitude of the fish farmers of the two districts. The findings are in-agreement with that of Dixit (1988), Mahandra Kumar (1996), Surekha *et al.* (1997) and Awasthi *et al.* (2008).

It is evident from the results that a positive and highly significant correlation exist between the innovative proneness of the fish farmers and their attitude towards scientific fish culture. The results imply that the higher the innovative proneness, the more favourable the attitude towards the enterprise. Innovative proneness is the sign of a lively mind ever seeking ways of making life and its activities meaningful and rich in quality. It is product of confidence and conviction, for which a favourable attitude is pre-requisite. Similar findings are reported by Nagarajaiah (2002).

The results indicate a non-significant relationship of credit orientation with the attitude of fish farmers towards scientific fish culture. This implies that irrespective of credit orientation, the farmers prefer their own attitude as already formed and this variable have no role in developing favourable attitude towards the fish culture. The findings are in line with the results reported by Ponnappan (1982), Bhakhri (1988) and Sunder Rao (1988).

A positive and highly significant relationship is found between value orientation and the attitude of the fish farmers of the two districts towards scientific fish culture. It may be inferred that value orientation has influence on attitude formation. The findings of the study are in conformity with the findings reported by Dana (1987).

The results of the study reveal a positive and highly significant relationship between risk orientation and the attitude of the respondents of the two districts towards scientific fish culture. A fish farmer convinced about the advantages of new technology will take risk and may develop a favourable attitude towards it. The results imply that the higher the risk orientation, the more favourable the attitude. The findings are in agreement with the results of the study by reported by Prasad (1995), Mahandra Kumar (1996), Awasthi *et al.* (2000), Balasubramaniam *et al.* (2001) and Nagarajaiah (2002). However, the results reported by Siddaramaiah and Reddy (1993) differ from the findings of the present study.

A positive and highly significant association is found between economic motivation and the attitude of the fish farmers of the two districts towards scientific fish culture. Trained respondents with a higher economic motivation have wider exposure to various sources of information and extension services which enable them to know more about the improved technologies. The process of gaining knowledge about scientific fish farming coupled with the strengthening of the sprit by economic motivation leads to the development of a favourable attitude towards scientific fish culture. The findings of the study are in conformity with the findings of Ponnappan (1982), Perumal *et al.* (1988), Prasad (1995), Mahandra Kumar (1996), Awasthi *et al.* (2000) and Nagarajaiah (2000).

Knowledge exhibits positive and significant relationship with the attitude of fish farmers of the two districts. The results imply that respondents with more knowledge have more favourable attitude towards scientific fish culture. To conclude, the farmers have a high level of knowledge and favourable attitude on the whole. The above results are in conformity with the findings of Samasundram and Singh (1978), Sidhu (1980) and Nagarajaiah (2002).

No significant relationship is noticed in North 24 Parganas between the size of the water body and the attitude of fish farmers towards scientific fish culture. The findings of the study are in contrast with the findings of Ponnappan (1982) and Mahandra Kumar (1996). However, these findings are in agreement with the findings reported by Nagarajaiah (2000). But in Dakshin Dinajpur a positive and highly significant correspondence is found between the size of water body and the attitude of fish farmers towards scientific fish culture.

No significant relationship is noticed between duration of water availability and attitude of the fish farmers towards scientific fish culture of the two districts. The findings of the study are in contrast with the findings of Nagarajaiah (2002).

No significant relationship is observed between the source of water and the attitude of the fish farmers in North 24 Parganas. But in Dakshin Dinajpur a positive and highly significant relationship is found between the source of water and the attitude of the fish farmers towards scientific fish culture. The findings in respect of Dakshin Dinajpur are in line with the findings of Nagarajaiah (2002).

No significant relationship is observed between the extent of weed infestation and the attitude of the fish farmers of North 24 Parganas. The extent of weed infestation has a positive and significant relationship with the attitude of the fish farmers of Dakshin Dinajpur. The findings of North 24 Parganas are in contrast with the findings of Nagarajaiah (2002).

2. Relationship between Socio-personal, Socio-economic, Communication, Psychological and Situational Variables of Fish Farmers and their Adoption Behaviour towards Scientific Fish Culture Practices

Simple correlation coefficients are computed for determination of the relationship between the independent variables and the adoption level of the fish farmers. The results are presented in Table 33.

North 24 Parganas

It is observed from the Table 33 that out of the 24 independent variables studied, 14 variables are significantly correlated with the adoption behaviour of the fish farmers towards scientific fish culture practices and 10 variables are found to have non-significant relationship.

Family size (X_3), family type (X_4), education (X_5), fish farming experience (X_6), annual income (X_8), social participation (X_{10}), possession of fishing equipment

(X_{11}), credit orientation (X_{16}), extent of weed infestation (X_{24}) have a non significant relationship with the adoption behaviour of the fish farmers towards scientific fish culture practices.

Table 33: Correlation of different Independent Variables with the Adoption Behaviour of Fish Farmers towards Scientific Fish Culture as Dependent Variable in North 24 Parganas (N=120) and Dakshin Dinajpur (N=120) District

Independent Variables	*Correlation Coefficient (r)*	
	Pearson Coefficient	
	North 24 Parganas (N=120)	*Dakshin Dinajpur (N=120)*
Age	–0.190*	–0.229 *
Caste	–0.212*	0.0089 NS
Family size	–0.159 NS	0.027 NS
Family type	–0.170 NS	–0.1158 NS
Education	0.063 NS	0.143 NS
Fish farming experiences	0.124 NS	0.225*
Occupation	0.368 **	0.123 NS
Annual income	0.145 NS	0.243 **
Land holding	0.269 **	0.251 **
Social participation	0.146 NS	0.316 **
Possession of fishing equipment	0.168 NS	0.149 NS
Mass media participation	0.370 **	0.490 **
Extension agency contact	0.383 **	0.620 **
Cosmopoliteness	0.384 **	0.478 **
Innovative proneness	0.546 **	0.554 **
Credit orientation	–0.047 NS	0.060 NS
Value orientation	0.385 **	0.512 **
Risk orientation	0.277 **	0.449 **
Economic motivation	0.468 **	0.549 **
Knowledge	0.505 **	0.736 **
Size of water body	0.241 **	0.271 **
Duration of water availability	0.215*	–0.006 NS
Source of water	–0.155 NS	0.075 NS
Extent of weed infestation	–0.025 NS	0.279 **

NS = Non significant; * = Significant at 0.05 level of probability; ** = Significant at 0.01 level of probability

Age (X_1), caste (X_2) and duration of water availability (X_{22}) have a significant relationship at 5 per cent level while some other variables such as occupation (X_7), land holding (X_9), mass media participation (X_{12}), extension agency contact (X_{13}), cosmopoliteness (X_{14}), innovative proneness (X_{15}), value orientation (X_{17}), risk orientation (X_{18}), economic motivation (X_{19}), knowledge about scientific fish culture practices (X_{20}) size of water body (X_{21}) show a highly significant and positive relationship with the adoption behaviour of the fish farmers about scientific fish culture practices at 1 per cent level.

The variables which have a negative relationship with the adoption behaviour of the fish farmers include age (X_1), caste (X_2), family size (X_3), family type (X_4), credit orientation (X_{16}), source of water (X_{23}) and the extent of weed infestation (X_{24}).

Dakshin Dinajpur

It is observed from the table that out of the 24 independent variables studied, 15 variables are positively and significantly correlated with the adoption behaviour of fish farmers towards scientific fish culture practices and the remaining 09 variables are found to have a non significant relationship.

It is clear from the table that most of the socio-personal and situational variables show a non significant relationship. The variables which show a non-significant relationship include caste (X_2), family size (X_3), family type (X_4), education (X_5), occupation (X_7), possession of fishing equipment (X_{11}), credit orientation (X_{16}), duration of water availability (X_{22}) and source of water (X_{23}).

The findings of the study further indicate that age (X_1) fish farming experience (X_6) has a significant relationship at 5 per cent level while the variables like annual income (X_8), land holding (X_9), social participation (X_{10}), mass media participation (X_{12}), extension agency contact (X_{13}), cosmopoliteness (X_{14}), innovative proneness (X_{15}), value orientation (X_{17}), risk orientation (X_{18}), economic motivation (X_{19}), knowledge about scientific fish culture (X_{20}), size of water body (X_{21}) and extent of weed infestation (X_{24}) have a highly significant relationship with the adoption behaviour of the fish farmers at 1 per cent level.

Among the variables which have negative relationship with the adoption behaviour of the fish farmers are age (X_1), family type (X_4), and duration of water availability (X_{22}).

In the two districts, age is found to have a negative as well as significant relationship with the adoption level of fish farmers. It is obvious that younger farmers being more enthusiastic are more easily attracted to scientific technologies. Hence the younger respondents adopted more practices. The findings of the study are in agreement with the findings reported by Sujath Kumar (1988), Das *et al.* (1988) and Talukdar and Sontaki (2005). But in Dakshin Dinajpur age and adoption level of fish farmers exhibit a non-significant relationship.

Caste is found to have a negative and significant relationship with adoption level of fish farmers in North 24 Parganas. The results imply that backward castes adopt more of the scientific fish culture practices. Caste and adoption level of fish farmers exhibit non-significant relationship in Dakshin Dinajpur. It reveals that irrespective of caste influence the adoption of scientific fish culture has its own way. Similar finding are reported by Subhas Chandra (1986).

Family size and adoption level of fish farmers exhibit a non-significant relationship in the two districts. The findings are in conformity with the results of

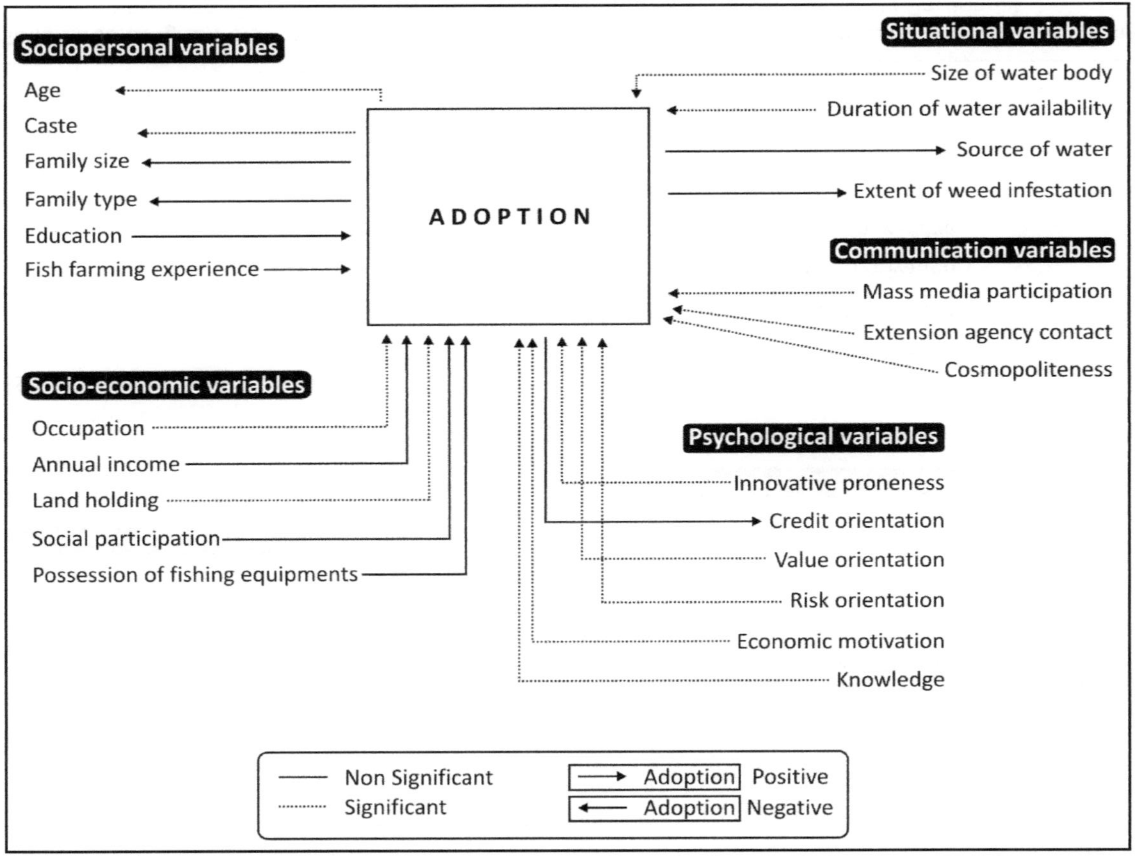

Figure 23: Schematic Diagram Showing the Empirical Relationship between Attitude of Fish Farmers towards Scientific Fish Culture Practices and Independent Variables (Dakshin Dinajpur).

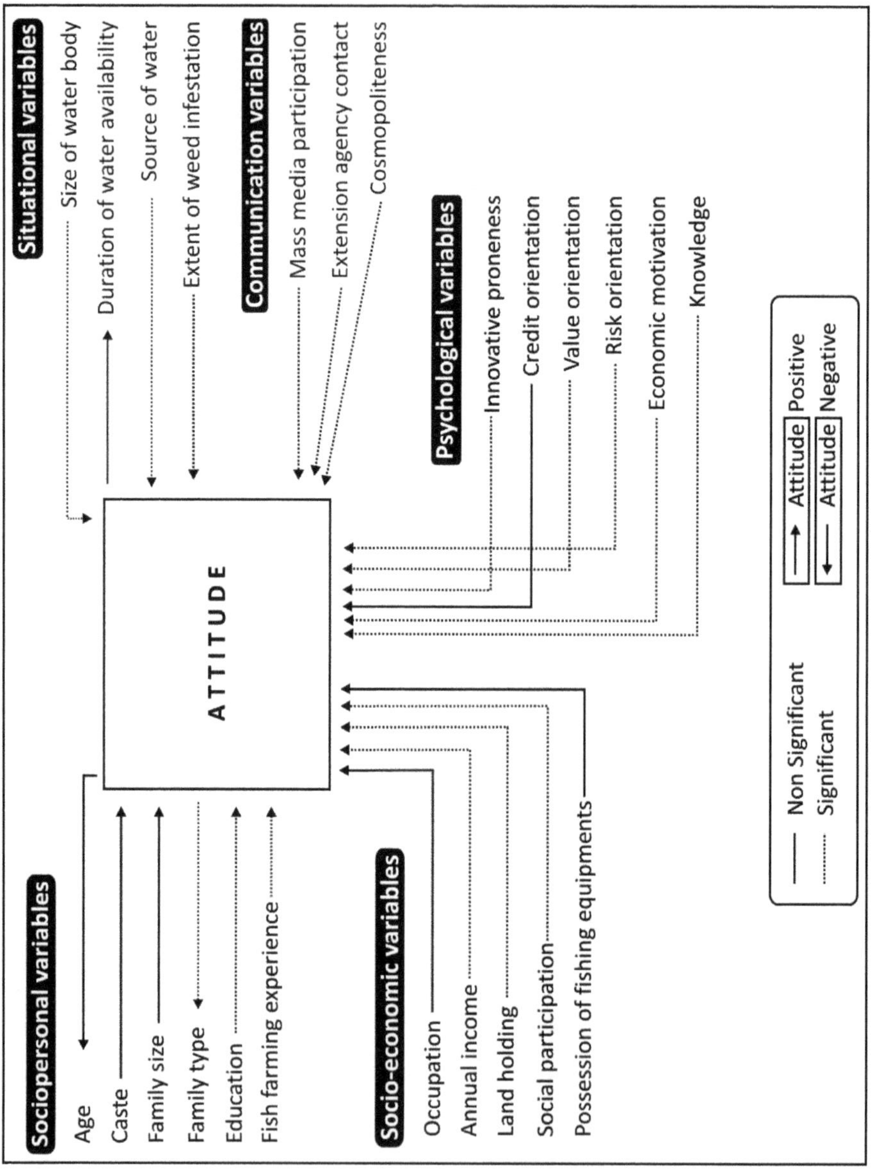

Figure 24: Schematic Diagram Showing the Empirical Relationship between Adoption Behaviour of Fish Farmers towards Scientific Fish Culture Practices and Independent Variables (North 24 Parganas).

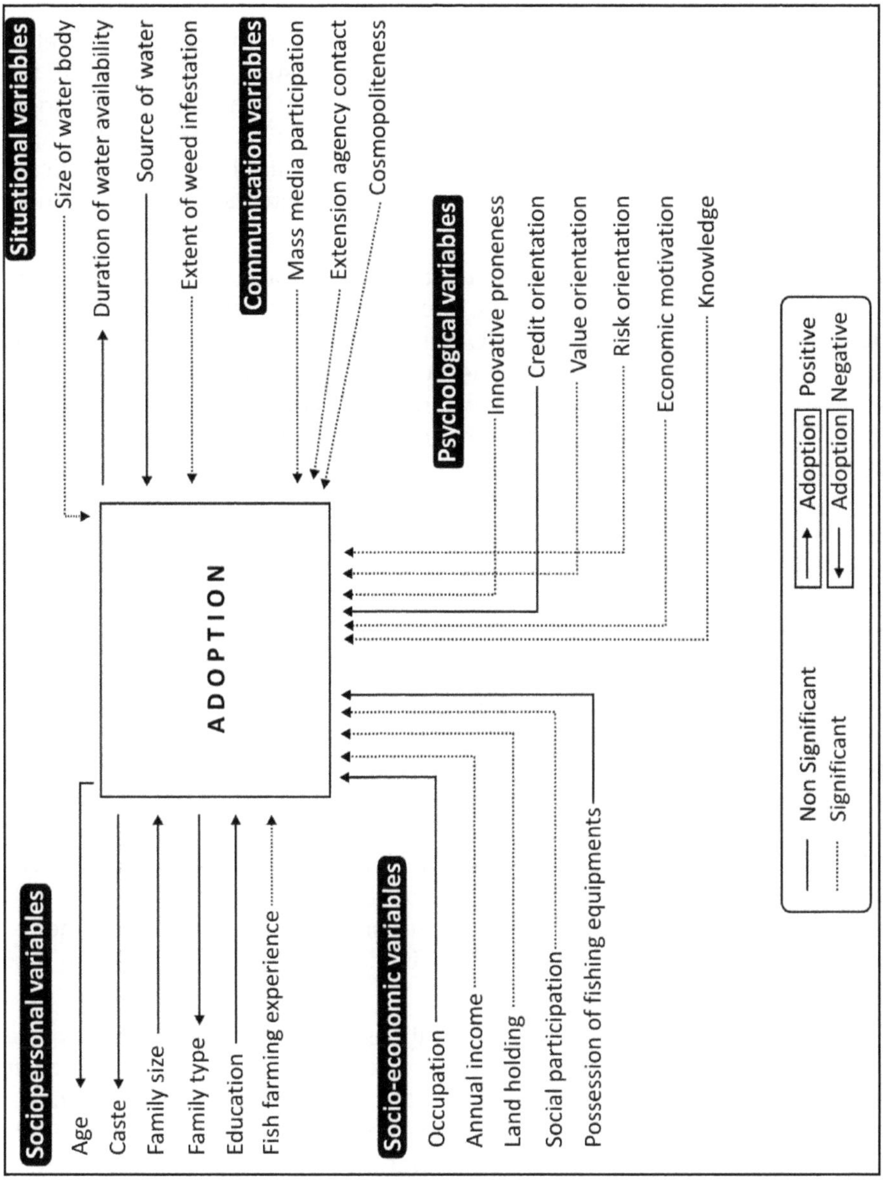

Figure 25: Schematic Diagram Showing the Empirical Relationship between Adoption Behaviour of Fish Farmers towards Scientific Fish Culture Practices and Independent Variables (Dakshin Dinajpur).

Reddy (1997). The results imply that the number of members in a family does not influence the adoption level.

Non-significant and negative relationship exist between family type and adoption level of fish farmers of the two districts. Similar findings are reported by Reddy (1997) and Nagarajaiah (2002).

Non-significant relationship exist between education and adoption level of the fish farmers of the two districts. The findings of the study are in line with the findings reported by Balasubramaniam (1988), Sheoran and Ramkumar (1988), Ingle *et al.* (1991), Nagarajaiah (2002) and Talukdar and Sontaki (2005). The results imply that as the education level of fish farmers is low, it can not have any significant impact on adoption behavior. Irrespective of education attained, the fish farmers adopt the recommended practices.

The results indicate a non-significant relationship in North 24 Parganas between fish farming experiences and adoption level of scientific fish culture practices. Practice of scientific fish culture in ponds being a subsidiary enterprise, new ideas of fish farming experience do not influence its adoption. Fish farmers have to know more, the latest information can enhance their knowledge and subsequent adoption. The findings are in line with the findings reported by Subhas Chandra (1986), Mathiyalagan and Subramaniam (1995), Mathiyalagan (1997), Subashini and Thyagarajan (2000), Nagarajaiah (2002) and Talukdar and Sontaki (2005). But in Dakshin Dinajpur a positive and significant relationship exist between the fish farming experience and adoption of scientific fish culture practices. The results imply that length of fish farming experience influences its adoption.

The results reveal that occupation have a positive and significant relationship with the adoption level of the fish farmers of North 24 Parganas. Respondents who pursue fish culture as primary occupation for supporting the family are better adopters, as they are completely dependent on fishery for their livelihood. The results imply that more effort should be put in by extension personnel for improving the adoption of the technology by those who pursue fish farming as a secondary occupation. These findings are in agreement with the findings reported by Mahandra Kumar (1995). But in Dakshin Dinajpur a non-significant and negative relationship exist between the variety of occupation and the adoption level of fish farmers. Respondents who pursue fish culture as primary occupation need to adopt more scientific fish culture practices as profitability, high economic return and comparative simplicity of the practice.

The results indicate a non-significant relationship between the annual income and the adoption level of scientific fish culture practices in North 24 Parganas. It implies that annual income has no influence on the adoption level of the fish farmers. The findings are in agreement with the findings reported by Mohammed Rafi (1993) and Shivasankarappa (1995). A positive and significant relationship exist between the annual income and the adoption of scientific fish culture practices in Dakshin Dinajpur. Greater awareness of the respondents will help the adoption of improved practices. Respondents with a higher income do not hesitate to increase investment on various inputs and management aspects of fish culture. With a view of realising more profit one has to adopt improved practices. This might be the plausible reason for a significant relationship between the annual income and the adoption level of fish farmers. The findings agree with the findings reported by Chede (1988), Sujath Kumar (1988), Mahandra Kumar (1996), Mathiyalagan (1997), Shinde *et al.* (1999), Sah and Ramchand (1999) and Nagarajaiah (2002).

The results reveal that there is significant relationship between the size of land holding and the adoption level of fish farmers of the two districts. The findings of Sheoran and Ram Kumar (1988), Chede (1988), Reddy (1997), Shinde *et al.* (1999), Sah and Ramchand (1999), Nagaraj *et al.* (2000) and Venkatesh Prasad and Siddaramaiah (2000) corroborate the results of the present study.

The results indicate a non-significant relationship in North 24 Parganas between social participation and adoption level of scientific fish culture practices. It implies that social participation has no influence on the adoption level of the fish farmers. Social participation has a positive and significant relationship with the adoption level of the fish farmers of Dakshin Dinajpur. The reason may be that the respondents may have come across and imbibed new ideas related to scientific fish culture practices while participating in the activities of different social organizations. Present findings are in line with the findings of Meeran (1983), Sujath Kumar (1988), Srishnaiah (1989), Mahandra Kumar (1996), Reddy (1997), Nagaraj *et al.* (2000), Venkatesh Prasad and Siddaramaiah (2000) and Nagarajaiah (2002).

The results reveal that there is no significant relationship between the possession of fishing equipments and the adoption level of the fish farmers of the two districts. It implies that the possession of fishing equipments has no the influence on the adoption level of the fish farmers. The findings of Mathiyalagan and Subramanian (1995) and Mathiyalan (1997) are in conformity with the findings of this study.

Mass media participation of the fish farmers of the two districts has a positive and significant correlation with the adoption of scientific fish culture practices. Fish farmers gain more knowledge related to fish culture practices when exposed to different mass media sources. It can help them to broaden their knowledge base. It also gives them a chance to learn about the useful role of the various associated benefits like training, credit and subsidy, etc. These findings are in conformity with the results of Sujath Kumar (1988), Das *et al.* (1988), Mathiyalan and Subramanian (1995), Mathiyalagan (1997), Nagaraj (1999), Nagaraj *et al.* (2000), Meeran (2000) and Nagarajaiah (2002).

A positive and highly significant correlation exist in the two districts between extension agency contact and adoption of scientific fish culture practices. It implies that the adoption level of the respondents is significantly influenced by their contacts with the extension agencies. By contact with the extension workers, fish farmers gain knowledge, and ultimately get convinced about the advantages of adopting improved practices. This helps increased adoption. The findings are in agreement with the findings reported by Meeran (1983), Subhas Chandra (1986), Balasubramaniam (1988), Bhaumik *et al.* (1992), Shinde *et al.* (1999), Nagaraj *et al.* (2000), Meeran (2000), Subashini and Thyagarajan (2000) and Nagarajaiah (2002).

A positive and highly significant relationship exist between the degrees of cosmopoliteness of the fish farmers of the two districts and extent of the adoption of scientific fish culture practices. The probable reason for the significant relationship is that the farmers who come in contact with the social system outside have better opportunities to exchange ideas and experience and have an useful interaction with other people. Thus, an individual having a higher cosmopoliteness, manages an enterprise in a cost effective manner with an eye to profit maximization. It implies that those farmers with higher contacts beyond their own social system have more adoption incentive. The findings of the present study get support from the works of Mahandra Kumar (1996), Reddy (1997), Shinde *et al.* (1999), Venkatesh Prasad and Siddaramaiah (2000), Chandrakala and Eswarappa (2001), Nagarajaiah (2002) and Talukdar and Sontaki (2005).

In the two districts innovative proneness has a positive and highly significant relationship with the adoption level of scientific fish culture practices. Innovative proneness is related to the willingness of an individual to accept the changes leading to the adoption of innovative ideas and practices of scientific fish culture. An individual with a high innovative proneness has generally a higher orientation towards risks. Scientific innovation and competition in the market economy influence the extent of his adoption of scientific fish culture practices. Similar findings are reported by Moulik (1965), Biswas *et al.* (1991) and Nagarajaiah (2002).

Non significant relationship exist between credit orientation and adoption level of the fish farmers of the two districts. It reveals that irrespective of credit orientation, fish farmers adopt scientific fish culture practices. As most of the fish farmers are economically not sound, they have to obtain credit for procurement of necessary inputs. It is observed that the greater number of respondents obtain credit from money lenders, middle men and fish wholesaler and retailers as most of the commercial banks do not consider fish culture as a commercially viable or profitable enough an enterprise for advancing credit. Present findings are in line with the findings of Sujath Kumar (1988), Mathiyalagan and Subramanian (1995), Mathiyalagan (1997) and Nagaraj (1999).

Value orientation has a positive and highly significant relationship with the adoption level of the fish farmers of the two districts. It implies that the adoption level of the fish farmers is significantly influenced by value orientation.

The results reveal that there is a positive and highly significant relationship between risk orientation and the adoption level of the fish farmers of the two districts. The findings are in conformity with the findings of Biswas *et al.* (1991), Krishnamurthy (1997), Nagaraj (1999), Meeran (2000) and Nagarajaiah (2002). It can be concluded that the respondents having a higher level of risk orientation are better adopters. It is rather interesting to note that fish farmers adopt scientific fish culture practices inspite of the possible risk of the enterprise.

The results indicate a positive and highly significant relationship between economic motivation and the adoption level of the fish farmers of the two districts. It implies that those farmers who have a tendency to maximize their earning and strive towards this end have higher adoption level. Economic motivation is the basic component upon which other motives and other incentives are dependent and which psychologically induces an individual to reorient himself to achieve higher income. Scientific fish culture being an economic and profit earning enterprise, one is likely to have a higher level of economic motivation and try to achieve the goal of higher income. The fish farmer would naturally be motivated hard to work for achieving maximum profits through proper adoption of scientific practices. Hence a significant relationship between economic motivation and extent of adoption is expected. The findings are in conformity with the findings of Biswas *et al.* (1991), Nagarajaiah (2002) and Talukdar and Sontaki (2005).

The knowledge level of the respondents have a positive and significant relationship with the adoption behaviour of the fish farmers of the two districts. A fish farmer possessing knowledge is more prone to adopt scientific fish farming practices. The combination of scientific and economic orientation enhances the prospects of resulting in the success of an enterprise. The findings are in accordance with the findings reported by Talukdar and Sontaki (2005).

The size of the water body (fish culture pond) has a significant relationship with the adoption behaviour of the fish farmers of the two districts. It implies that the size of the water body has a significant role in promoting the adoption of scientific fish farming practices as the variable is supposed to be an important factor in influencing the adoption behaviour. The present findings are in conformity with the findings reported by Mishra (1987), Balasubramaniam (1988), Das *et al.* (1987), Krishnaiah (1989), Srivastava *et al.* (1993), Mahandra Kumar (1996). However, findings reported by Subhas Chandra (1986), Nagarajaiah (2002) and Talukdar and Sontaki (2005) differ from the present findings.

The results reveal that there is a positive and significant relationship between the duration of water availability and the adoption level of the fish farmers of North 24 Parganas. The probable reason is that the fish farmers possessing long seasonal ponds could take up scientific fish culture more systematically and obtain better production. Duration of water availability exhibit a non-significant relationship with the extent of adoption by the fish farmers of Dakshin Dinajpur. The results indicate that the adoption level is not influenced by the duration of water availability. The probable reason for this is that the respondents have better knowledge of water availability in ponds and adopt the practices. The findings are in agreement with the findings reported by Balasubramaniam (1988) and Nagarajaiah (2002).

A non-significant relationship is found to exist between the source of water and the adoption level of the fish farmers of the two districts. Since the fluctuation in the variable is comparatively negligible, the non-significant relationship is justified. The findings are in accordance with findings reported by Balasubramaniam (1988), Nagarajaiah (2002).

A non-significant relationship is found to exist between the extent of weed infestation and the adoption level of the fish farmers of North 24 Parganas. The results also indicate that a positive significant relationship is exhibited between the extent of weed infestation and the adoption level of fish culture practices in Dakshin Dinajpur. This implies that the respondents possessing ponds with a quantity of aquatic weeds that may be termed are better adopters. Hence, it can be concluded that aquatic weeds management requires to be focussed by extension personnel, help being made available to the farmers by means of technical and financial support. The findings are in conformity with the findings of Nagarajaiah (2002).

B. Multiple Regression

1. Extent of Contribution of Independent Variables on the Attitude of the Fish Farmers towards Scientific Fish Culture Practices

The multiple regression analysis is undertaken to determine the extent of contribution of independent variables to the attitude of fish farmers towards scientific fish culture practices.

North 24 Parganas

It can be concluded that in North 24 Parganas, variables like caste, family size, annual income, cosmopoliteness, innovative proneness, knowledge and size of the water body significantly influence the variation in the attitude of the fish farmers and these seven variables influence the attitude of the farmers can be termed as good predictors of attitude of the fish farmers towards scientific fish culture in North 24 Parganas. As against this in Dakshin Dinajpur, the variables like education, annual income and social participation significantly influence the variation in the attitude of fish farmers, and these three variables influencing attitude of farmers can be termed as good predictors of the attitude of the fish farmers and multiplication of these variables can help in further influencing the attitude of the fish farmers towards scientific fish culture practices in Dakshin Dinajpur. A significant change in the attitude of fish farmers can be effected through manipulating, and bringing positive changes in, these variables, thereby improving their production levels with increased adoption.

Table 34: Stepwise Multiple Regression of Attitude of Fish Farmers towards Scientific Fish Culture Practices with Independent Variables in North 24 Parganas

Independent Variables	*Unstandardize Coefficients*		*Standardize Coefficient*	*t*	*Rank*
	B	*SE*	*Beta*		
Caste	–1.932	0.609	–0.213	–3.173**	**VI**
Family size	2.050	0.738	0.179	2.776**	**IV**
Annual income	2.255	0.626	0.275	3.601**	**II**
Cosmopoliteness	0.653	0.286	0.195	2.282*	**III**
Innovative proneness	0.299	0.167	0.132	1.790	V
Knowledge	0.254	0.053	0.394	4.764**	**I**
Size of water body	–1.817	0.540	–0.248	–3.362**	**VII**

R^2 = 0.56 SE(est)= 3.94 * Significant at 0.05 level of probability; ** Significant at 0.01 level of probability.

Dakshin Dinajpur

A close observation of Table 35 reveals that three variables *viz.*, education, annual income and social participation are significant in multiple regression analysis in explaining the variations in the attitude of the fish farmers. Further it may be observed from the table that 27 per cent of the variations in the attitude of the fish farmers towards scientific fish culture practices are due to the combined influence of the selected variables included in the analysis. R^2 value of 0.27 reveals the significant regression at 5 per cent or 1 per cent level. The independent variables are ranked on the basis of standard Beta values to find out their relative importance in predicting the dependent variable. Annual income, social participation and education occupy the first three ranks and they have relative importance in predicting the attitude of the fish farmers. These three variables could be termed as good predictors of the attitude of the fish farmers. These variables help in influencing the attitude towards the adoption of scientific fish culture.

Table 35: Stepwise Multiple Regression of Attitude of Fish Farmers towards Scientific Fish Culture Practices with Independent Variables in Dakshin Dinajpur

Independent Variables	*Unstandardize Coefficients*		*Standardize Coefficient*	*t*	*Rank*
	B	*SE*	*Beta*		
Education	1.14	0.57	0.17	2.00*	III
Annual income	3.08	1.12	0.26	2.75**	I
Social participation	0.76	0.32	0.23	2.39*	II

R^2 = 0.27, SE (est)= 7.38 * Significant at 0.05 level of probability; ** Significant at 0.01 level of probability

It could be concluded that in North 24 Parganas, the variables *viz.*, caste, family size, annual income, cosmopoliteness, innovative proneness, knowledge and size of the water body significantly influence the variations in the attitude of the fish farmers and these seven variables influence the attitude of the farmers can be termed as good predictors of the attitude of the fish farmers towards scientific fish culture in North 24 Parganas while in Dakshin Dinajpur, the variables like education, annual income and social participation significantly influence the variations in the attitude of the fish farmers, and these three variables influencing the attitude of the farmers can be termed as good predictors of the attitude of the fish farmers and the multiplication of these variables could help in further influencing the attitude of the fish farmers towards scientific fish culture practices in Dakshin Dinajpur. Significant changes in the attitude of fish farmers can be effected through manipulating, and bringing positive changes in, these variables, thereby improving their production levels with increased adoption.

2. Contribution of Independent Variables to the Adoption Behaviour of Fish Farmers towards Scientific Fish Culture Practices

The multiple regression analysis is undertaken to determine the extent of the contribution of the independent variables to the adoption behaviour of the fish farmers towards scientific fish culture practices.

North 24 Parganas

A close observation of Table 36 reveals that ten variables *viz.*, age, family type, occupation, land holding, social participation, extension agency contact, innovative proneness, credit orientation, knowledge and source of water are significant in multiple regression analysis in explaining the variations in the adoption level of fish farmers.

Further it may be observed from table that 71 per cent of the variation in the adoption behaviour of the fish farmers towards scientific fish culture practices could be explained by variables included in the study. R^2 value of 0.71 reveals significant regression at 5 per cent or 1 per cent level. The independent variables are ranked on the basis of standard Beta values to find out their relative importance in predicting the dependent variables. Occupation, extension agency contact and innovative proneness, land holding and Knowledge occupy the first five ranks, each with relative importance in predicting the adoption behaviour of the fish farmers. These ten variables could be termed as good predictors of the adoption behaviour of the fish farmers. These variables help in increasing the level of the adoption of scientific fish culture.

Table 36: Stepwise Multiple Regression of Adoption Behaviour of Fish Farmers towards Scientific Fish Culture Practices with some Independent Variables in North 24 Parganas

Independent Variables	*Unstandardize Coefficients*		*Standardize Coefficient*	*t*	*Rank*
	B	*SE*	*Beta*		
Age	–3.405	0.795	–0.261	–4.281**	X
Family type	–4.041	1.046	–0.237	–3.862**	IX
Occupation	7.872	1.085	0.408	7.255**	I
Land holding	1.795	0.457	0.227	3.931**	IV
Social participation	–0.572	0.213	–0.194	–2.678**	VIII
Extension agency contact	1.589	0.246	0.374	6.460**	II
Innovative proneness	1.133	0.195	0.340	5.815**	III
Credit orientation	–0.717	0.268	–0.149	–2.673**	VI
Knowledge	0.148	0.065	0.156	2.285*	V
Source of water	–4.025	1.273	–0.171	–3.163**	VII

R^2 =0.71, SE(est)= 4.75 * Significant at 0.05 level of probability; ** Significant at 0.01 level of probability

Dakshin Dinajpur

A close observation of Table 37 reveals that eight variables *viz.*, age, education, annual income, land holding, extension agency contact, innovative proneness, knowledge and extent of weed infestation are significant in multiple regression analysis in explaining the variation in the attitude of fish farmers.

Further, it may be observed from the Table that 73 per cent of the variations in the adoption behaviour of the fish farmers towards scientific fish culture practices could be explained by the variables included in the analysis. R^2 value of 0.73 reveals the significant of regression at 5 per cent or 1 per cent level. The independent variables are ranked on the basis of standard Beta values to find out their relative importance in predicting the dependent variable. Knowledge, extension agency contact and innovative proneness occupy the first three ranks and they have relative importance in predicting the adoption behaviour of the fish farmers. These eight variables could be termed as good predictors of the adoption behaviour of fish farmers. These variables help in increasing the level of adoption of scientific fish culture.

Table 37: Stepwise Multiple Regression of Adoption Behaviour of Fish Farmers towards Scientific Fish Culture Practices with some Independent Variables in Dakshin Dinajpur

Independent Variables	*Unstandardize Coefficients*		*Standardize Coefficient*	*t*	*Rank*
	B	*SE*	*Beta*		
Age	–1.95	0.90	–0.11	–2.16*	**VI**
Education	–1.47	0.53	–0.15	–2.75**	**VIII**
Annual income	–2.30	1.13	–0.13	–2.03*	**VII**
Land holding	2.12	0.78	0.16	2.70**	**V**
Extension agency contact	1.98	0.31	0.37	6.34**	**II**
Innovative proneness	1.02	0.25	0.24	4.01**	**III**
Knowledge	0.37	0.07	0.42	5.73**	**I**
Extent of weed infestation	2.68	0.79	0.18	3.40**	**IV**

R^2 =0.73, SE(est)= 6.70 Significant at 0.05 level of probability; ** Significant at 0.01 level of probability

In North 24 Parganas, among the selected variables ten variables *viz.*, age, family type, occupation, land holding, social participation, extension agency contact, innovative proneness, credit orientation, knowledge and source of water significantly contribute to the variation in the adoption level of the fish farmers. Hence, it could be concluded that occupation, extension agency contact, innovation proneness, land holding and knowledge positively and significantly influence the variations in the adoption behaviour of the fish farmers. These ten variables could be termed as good predictors of the fish farmers' level of adoption and the multiplication of these variables could help to increase the adoption of scientific fish culture practices. Whereas in Dakshin Dinajpur eight variables *viz.*, age, education, annual income, land holding, extension agency contact, innovative proneness, knowledge and extent of weed infestation significantly contribute to the variations in the adoption pattern of the fish farmers and the five variables, namely, knowledge extension agency contact, innovation proneness, extent of weed infestation and land holding are positively and significantly influential in bringing about the variations in the adoption behaviour of the fish farmers. Das *et al.* (1988)

report 55 per cent variation in the adoption of composite fish culture practices on account of 14 variables whereas Reddy (1997) reports 47 per cent variations in the adoption level on account of 21 variables.

C. Path Analysis

Direct and Indirect Effects of Independent Variables on the Attitude of Fish Farmers towards Scientific Fish Culture Practices

For arriving at a quantitative interpretation of the direct and indirect effects of selected variables included in the study and for developing the operational statistical mode, the path analysis, being an effective and appropriate device is employed. The results are presented in table and represented in the figure. Only those variables which are significantly correlated with the attitude of fish farmers are considered for path analysis.

North 24 Parganas

Direct Effects

A glance at the Table 38 and Figure 26 reveals that out of 23 variables included in the path analysis, 14 variables have a positive direct effect, and the remaining 9 variables alone have negative direct effect on the attitude of the fish farmers towards scientific fish culture.

The ranking of variables based on their direct effect on the attitude of the fish farmers reveals that knowledge about scientific fish culture (X_{20}), cosmopoliteness (X_{14}), economic motivation (X_{19}), occupation(X_7) and family type (X_4) occupy the first five ranks in that order whereas fish farming experience (X_6) risk orientation (X_{18}), extension agency contact (X_{13}), value orientation (X_{17}), size of water body (X_{21}) fall into the last five ranks in order.

Indirect Effects

The total indirect effects channelized through other factors for each of the independent factors are also found to be quite substantial. The ranking of these effects indicates that value orientation (X_{17}), social participation (X_{10}), innovative proneness (X_{15}), risk orientation (X_{18}) and extension agency contact (X_{13}) occupy the first five ranks and have the largest total positive indirect effects on the attitude of the fish farmers to scientific fish culture, whereas caste (X_2), family size (X_3), family type (X_4), annual income (X_8) and age (X_1) occupy the last five ranks exerting minimum indirect effects.

Substantial Indirect Effect

The first largest substantial positive effect on the attitude of the fish farmers is exerted by the variable value orientation (0.308) through cosmopoliteness. The second largest substantial positive indirect effect on the attitude of the fish farmers is exerted by value orientation (0.301) through knowledge. The third largest substantial positive indirect effect is exerted by economic motivation (0.235) through value orientation. Out of 69 substantial indirect effects, 15 pass through each of the two independent variables: value orientation (X_{17}) and knowledge (X_{20}), ten pass

Table 38: Path Coefficient Showing Direct, Indirect and Substantial Indirect Effects of Independent Variables on the Attitude of Fish Farmers towards Scientific Fish Culture Practices in North 24 Parganas

Variable Code	Variables	Correlation Coefficient	Direct Effects	Rank	Total Indirect Effect	Rank	Variables through which Substantial Indirect Effects are Channeled through		
							I	II	III
X_1	Age	–0.150	0.062	11	–0.212	23	0.085 (X_{20})	0.077 (X_3)	0.065 (X_4)
X_2	Caste	–0.170	–0.097	17	–0.073	19	0.105 (X_7)	0.093 (X_{14})	0.047 (X_{17})
X_3	Family size	0.130	0.228	6	–0.098	20	0.149 (X_4)	0.066 (X_{19})	0.048 (X_{21})
X_4	Family type	0.140	0.272	5	–0.132	21	0.125 (X_3)	0.058 (X_8)	0.057 (X_{17})
X_5	Education	0.180	0.110	9	0.070	12	0.154 (X_{14})	0.141 (X_{20})	0.101 (X_{17})
X_6	Fish farming experience	0.100	–0.165	19	0.265	8	0.121 (X_{20})	0.098 (X_{17})	0.078 (X_{19})
X_7	Occupation	0.140	0.291	4	–0.151	22	0.082 (X_{21})	0.077 (X_6)	0.070 (X_{20})
X_8	Annual income	0.350	0.205	7	0.145	10	0.171 (X_{20})	0.170 (X_{21})	0.168 (X_{17})
X_9	Size of land holding	0.150	0.153	8	–0.003	17	0.197 (X_{21})	0.161 (X_{20})	0.141 (X_{17})
X_{10}	Social participation	0.350	–0.123	18	0.473	2	0.216 (X_{20})	0.187 (X_{14})	0.148 (X_{17})
X_{12}	Mass media participation	0.330	0.029	13	0.301	6	0.271 (X_{20})	0.187 (X_{14})	0.161 (X_{17})
X_{13}	Extension agency contact	0.150	–0.201	21	0.351	5	0.215 (X_{14})	0.171 (X_{20})	0.124 (X_{17})
X_{14}	Cosmopoliteness	0.480	0.467	2	0.013	15	0.286 (X_{20})	0.222 (X_{17})	0.161 (X_{19})
X_{15}	Innovative proneness	0.420	0.002	14	0.418	3	0.224 (X_{14})	0.211 (X_{20})	0.128 (X_{19})
X_{16}	Credit orientation	–0.020	–0.032	15	0.012	16	0.067 (X_{17})	0.060 (X_{19})	0.057 (X_4)
X_{17}	Value orientation	0.330	–0.336	22	0.666	1	0.308 (X_{14})	0.301 (X_{20})	0.209 (X_{19})
X_{18}	Risk orientation	0.240	–0.174	20	0.414	4	0.233 (X_{14})	0.226 (X_{20})	0.197 (X_{19})
X_{19}	Economic motivation	0.320	0.299	3	0.021	13	0.252 (X_{14})	0.246 (X_{20})	0.235 (X_{17})
X_{20}	Knowledge	0.620	0.502	1	0.118	11	0.266 (X_{14})	0.202 (X_{17})	0.146 (X_{19})
X_{21}	Size of water body	–0.040	–0.340	23	0.300	7	0.155 (X_{17})	0.146 (X_{19})	0.103 (X_8)
X_{22}	Duration of water availability	0.010	0.037	12	–0.027	18	0.170 (X_{21})	0.094 (X_{17})	0.093 (X_3)
X_{23}	Source of water	0.090	0.073	10	0.017	14	0.055 (X_{20})	0.043 (X_4)	0.039 (X_8)
X_{24}	Extent of weed infestation	0.160	–0.094	16	0.254	9	0.122 (X_4)	0.072 (X_8)	0.061 (X_7)

Residual effect : 0.604.

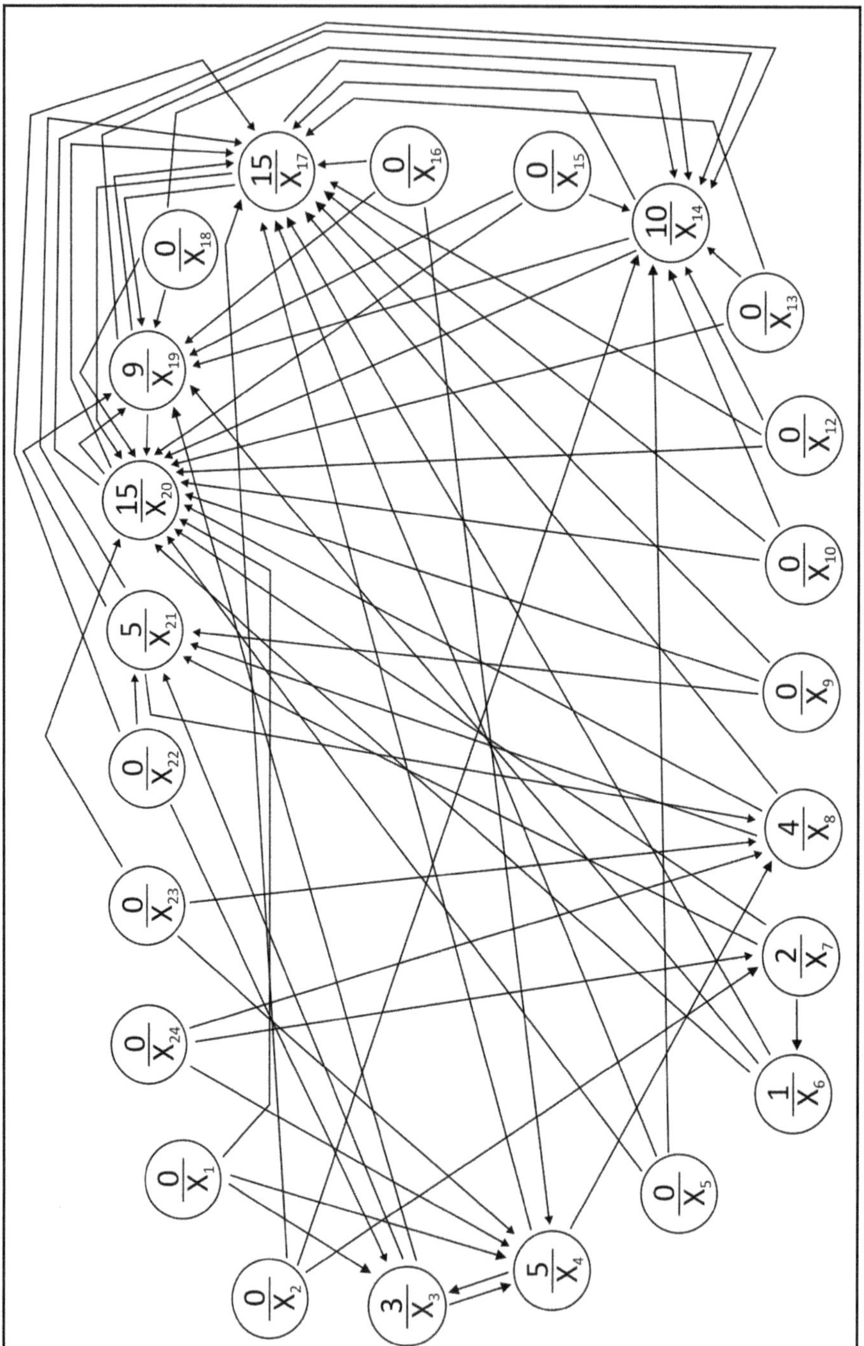

Figure 26: Diagrammatic Representation of the Variables through which Indirect Substantial Effects are Channeled (Attitude in North 24 Parganas).

through cosmopoliteness (X_{14}), nine pass through economic motivation (X_{19}), five pass through family type (X_4) and size of water body (X_{21}), four pass through annual income (X_8), three pass through family size (X_3), two pass through occupation (X_7) and one passes through fish farming experience (X_6). The residual effect is found to be 0.604 or in a way, 60 per cent of the total variability has been left unexplained.

Dakshin Dinajpur

Direct Effects

The Table 39 reveals that out of 20 variables included in the path analysis, 10 variables have a positive direct effect and the remaining 10 variables have a negative direct effect on the attitude of the fish farmers towards scientific fish culture practices.

The ranking of variables based on their direct effect reveals that economic motivation (X_{19}), knowledge about scientific fish culture (X_{20}), innovative proneness (X_{15}), extent of weed infestation (X_{24}) and cosmopoliteness (X_{14}) occupy the first five ranks in that order, whereas possession of fishing equipment (X_{11}), risk orientation (X_{18}), credit orientation (X_{16}), duration of water availability (X_{22}) and value orientation (X_{17}) obtain the last five ranks in the same order.

Indirect Effects

The total indirect effects channelized through other factors for each of the independent factors are also found to be quite substantial. The ranking of these effects indicates that value orientation (X_{17}), risk orientation (X_{18}), extension agency contact (X_{13}), cosmopoliteness (X_{14}), mass media participation (X_{12}) occupy the first five ranks which induce the highest total indirect effect on attitude of fish farmers in the descending order of magnitude whereas possession of fishing equipment (X_{11}), source of water(X_{23}), extent of weed infestation (X_{24}) Duration of water availability (X_{22}), age (X_1), occupy the last five ranks in order of exerting minimum indirect effect.

Substantial Indirect Effect

The first largest substantial indirect effect on the attitude of the fish farmers towards scientific fish culture is exerted by the variable value orientation (0.294) through economic motivation. The second largest substantial indirect effect is exerted by both value orientation and risk orientation (0.245) through knowledge. The third largest substantial indirect is exerted by value orientation (0.105) through innovative proneness. Out of 60 substantial indirect effects, 18 effects are routed through the variable knowledge (X_{20}), 17 effects through economic motivation (X_{19}), 10 pass through value orientation (X_{17}), 7 pass through innovative proneness (X_{15}), 5 pass through duration of water availability (X_{22}) and one each passes through fish farming experiences (X_6), cosmopoliteness (X_{14}) and credit orientation (X_{16}). The residual effect is found to be 0.504, or in a way 50 per cent of the total variability are left unexplained.

The key attribute of an attitude is its evaluative component. Hence it is very clear from the findings that in North 24 Parganas knowledge, value orientation, cosmopoliteness, economic motivation, family type, innovative proneness, annual income, family size and size of water body play an important role in developing the

Table 39: Path Coefficient Showing Direct, Indirect and Substantial Indirect Effects of Independent Variables on Attitude of Fish Farmers towards Scientific Fish Culture Practices in Dakshin Dinajpur

Variable Code	*Variables*	*Correlation Coefficient*	*Direct Effects*	*Rank*	*Total Indirect Effect*	*Rank*	*Variables through which Substantial Indirect Effects are Channeled through*		
							I	*II*	*III*
X_1	Age	–0.020	0.045	8	–0.065	20	0.062 (X_{20})	0.023 (X_6)	0.021 (X_{22})
X_5	Education	0.350	0.032	10	0.318	10	0.145 (X_{20})	0.081 (X_{19})	0.072 (X_{15})
X_6	Fish farming experiences	0.290	0.060	6	0.230	15	0.162 (X_{19})	0.129 (X_{20})	0.061 (X_{17})
X_8	Annual income	0.430	0.051	7	0.379	8	0.174 (X_{20})	0.153 (X_{19})	0.075 (X_{17})
X_9	Size of land holding	0.240	–0.028	15	0.268	13	0.149 (X_{20})	0.102 (X_{19})	0.054 (X_{22})
X_{10}	Social participation	0.430	–0.005	12	0.435	6	0.195 (X_{20})	0.128 (X_{19})	0.059 (X_{17})
X_{11}	Possession of fishing equipment	0.150	–0.061	16	0.211	16	0.087 (X_{20})	0.068 (X_{19})	0.056 (X_{22})
X_{12}	Mass media participation	0.440	–0.013	13	0.453	5	0.220 (X_{20})	0.162 (X_{19})	0.077 (X_{17})
X_{13}	Extension agency contact	0.470	–0.019	14	0.489	3	0.247 (X_{19})	0.199 (X_{20})	0.079 (X_{15})
X_{14}	Cosmopoliteness	0.620	0.133	5	0.487	4	0.264 (X_{19})	0.232 (X_{20})	0.088 (X_{17})
X_{15}	Innovative proneness	0.620	0.219	3	0.401	7	0.220 (X_{20})	0.192 (X_{19})	0.072 (X_{14})
X_{16}	Credit orientation	0.150	–0.119	18	0.269	12	0.128 (X_{19})	0.087 (X_{20})	0.034 (X_{22})
X_{17}	Value orientation	0.550	–0.138	20	0.688	1	0.294 (X_{19})	0.245 (X_{20})	0.105 (X_{15})
X_{18}	Risk orientation	0.510	–0.104	17	0.614	2	0.281 (X_{19})	0.245 (X_{20})	0.090 (X_{17})
X_{19}	Economic motivation	0.660	0.426	1	0.234	14	0.270 (X_{20})	0.099 (X_{15})	0.095 (X_{17})
X_{20}	Knowledge	0.770	0.415	2	0.355	9	0.277 (X_{19})	0.116 (X_{15})	0.082 (X_{17})
X_{21}	Size of water body	0.300	–0.003	11	0.303	11	0.196 (X_{19})	0.158 (X_{20})	0.071 (X_{17})
X_{22}	Duration of water availability	–0.020	–0.134	19	0.114	19	0.055 (X_{19})	0.037 (X_{17})	0.033 (X_{16})
X_{23}	Source of water	0.210	0.044	9	0.166	17	0.083 (X_{20})	0.038 (X_{22})	0.035 (X_{15})
X_{24}	Extent of weed infestation	0.280	0.161	4	0.119	18	0.122 (X_{20})	0.034 (X_{19})	0.026 (X_{15})

Residual effect : 0.504.

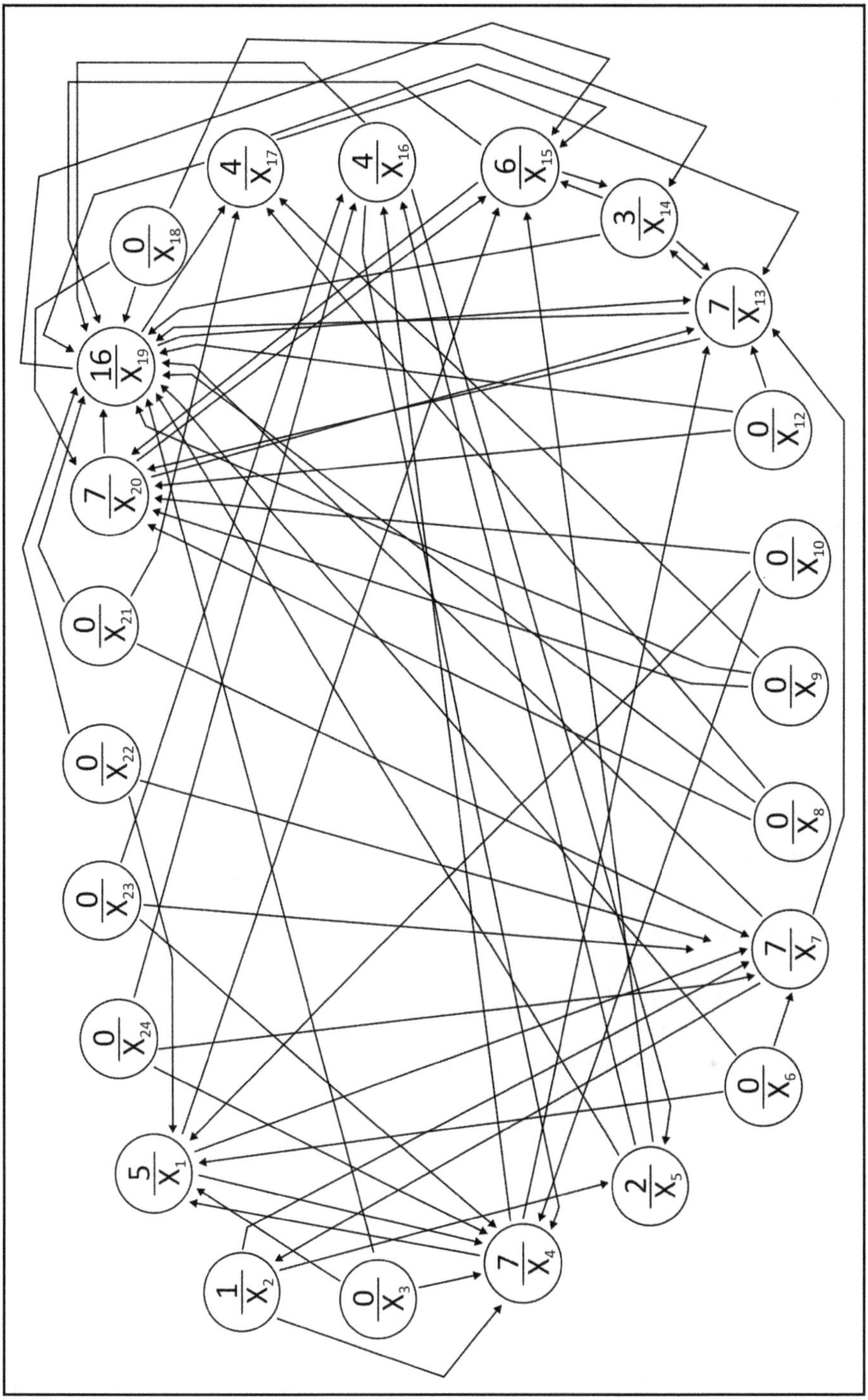

Figure 27: Diagrammatic Representation of the Variables through which Indirect Substantial Effects are Channeled (Attitude in Dakshin Dinajpur).

attitude of the respondents towards scientific fish culture practices. Any positive change in these variables would lead to the development of favourable attitude amongst the fish farmers. These results are in strongly corroborated by the findings of step wise multiple regression analysis.

In Dakshin Dinajpur knowledge, economic motivation value orientation, innovative proneness, cosmopoliteness, risk orientation and duration of water availability are found to be the dominant factors not only with direct effects on the attitude of the fish farmers but also with side impact on most of the other factors to influence the attitude of the fish farmers indirectly.

Direct and Indirect Effects of Independent Variables on Adoption Behaviour of Fish Farmer towards Scientific Fish Culture Practices

In order to know the direct and indirect effects, path analysis is carried out by considering the variables which are significantly correlated with the level of the adoption of scientific fish culture practices.

North 24 Parganas

Direct Effects

Table 40 reveals that out of 14 variables included in the path analysis, 8 variables have a positive direct effect and the remaining 6 variables have negative direct effect on the level of the adoption of scientific fish culture practices.

The ranking of variables based on their direct effect reveals that extension agency contact (X_{13}), innovative proneness (X_{15}), occupation (X_7), economic motivation (X_{19}), knowledge about scientific fish culture (X_{20}) occupy the first five ranks in order, whereas cosmopoliteness (X_{14}), age (X_1), source of water (X_{20}), credit orientation (X_{16}) and family type (X_4) obtain the last five ranks in the same order.

Indirect Effects

The total indirect effects channelized through other factors for each of the independent factors, are also found to be quite substantial. The ranking of these effects indicate that cosmopoliteness (X_{14}), value orientation (X_{17}), risk orientation (X_{18}), mass media participation (X_{12}), size of water body (X_{21}) occupy the first five ranks which have the highest total indirect effect on adoption in the descending order of magnitude, whereas occupation (X_7), extension agency contact (X_{13}), extent of weed infestation (X_{24}), family size (X_3) and caste (X_2) occupy the last five ranks in that order in exerting minimum indirect effect.

Substantial Indirect Effect

The first largest substantial indirect effect on the adoption behaviour of the fish farmers is exerted by the variable value orientation (0.214) through economic motivation. The second largest substantial indirect effect is exerted by cosmopoliteness (0.189) through extension agency contact. The third largest substantial indirect is exerted by cosmopoliteness through economic motivation. Out of 69 substantial indirect effects, 16 effects are routed through the variable economic motivation (X_{19}), 7 effects each through family type (X_4), occupation (X_7),

Table 40: Path Coefficient Showing Direct, Indirect and Substantial Indirect Effects of Independent Variables on Adoption Behaviour of Fish Farmers towards Scientific Fish Culture Practices in North 24 Parganas

Variable Code	*Variables*	*Correlation Coefficient*	*Direct Effects*	*Rank*	*Total Indirect Effect*	*Rank*	*Variables through which Substantial Indirect Effects are Channeled through*		
							I	*II*	*III*
X_1	Age	–0.190	–0.224	20	0.034	17	0.091 (X_4)	0.0988 (X_{15})	0.050 (X_7)
X_2	Caste	–0.210	0.115	7	–0.325	23	0.129 (X_7)	0.075 (X_5)	0.049 (X_4)
X_3	Family size	–0.160	0.063	11	–0.223	22	0.209 (X_4)	0.076 (X_1)	0.067 (X_{19})
X_4	Family type	–0.170	–0.379	23	0.209	10	0.082 (X_{13})	0.058 (X_{16})	0.054 (X_1)
X_5	Education	0.060	–0.192	18	0.252	8	0.108 (X_{15})	0.089 (X_{16})	0.076 (X_{19})
X_6	Fish farming experiences	0.120	–0.045	14	0.165	13	0.169 (X_7)	0.080 (X_{19})	0.076 (X_1)
X_7	Occupation	0.370	0.359	3	0.011	19	0.064 (X_{19})	0.058 (X_{13})	0.041 (X_2)
X_8	Annual income	0.150	0.119	6	0.031	18	0.119 (X_{19})	0.086 (X_{17})	0.083 (X_{20})
X_9	Size of land holding	0.270	0.111	8	0.159	14	0.092 (X_{19})	0.078 (X_{20})	0.072 (X_{17})
X_{10}	Social participation	0.150	–0.116	16	0.266	7	0.148 (X_4)	0.105 (X_{20})	0.089 (X_1)
X_{12}	Mass media participation	0.370	0.068	10	0.303	4	0.205 (X_{13})	0.131 (X_{20})	0.116 (X_{19})
X_{13}	Extension agency contact	0.380	0.411	1	–0.031	20	0.098 (X_{19})	0.094 (X_{14})	0.083 (X_{20})
X_{14}	Cosmopoliteness	0.380	–0.204	19	0.584	1	0.192 (X_{15})	0.189 (X_{13})	0.165 (X_{19})
X_{15}	Innovative proneness	0.550	0.399	2	0.151	15	0.131 (X_{19})	0.102 (X_{20})	0.098 (X_{14})
X_{16}	Credit orientation	–0.050	–0.277	22	0.227	9	0.080 (X_4)	0.062 (X_5)	0.061 (X_{19})
X_{17}	Value orientation	0.390	–0.172	17	0.562	2	0.214 (X_{19})	0.152 (X_{13})	0.148 (X_{15})
X_{18}	Risk orientation	0.280	–0.093	15	0.373	3	0.202 (X_{19})	0.110 (X_{20})	0.102 (X_{14})
X_{19}	Economic motivation	0.470	0.306	4	0.164	12	0.172 (X_{15})	0.131 (X_{13})	0.120 (X_{17})
X_{20}	Knowledge	0.510	0.243	5	0.267	6	0.168 (X_{15})	0.150 (X_{19})	0.140 (X_{13})
X_{21}	Size of water body	0.240	–0.042	13	0.282	5	0.150 (X_{19})	0.086 (X_7)	0.079 (X_{17})
X_{22}	Duration of water availability	0.220	0.030	12	0.190	11	0.085 (X_1)	0.070 (X_{19})	0.054 (X_7)
X_{23}	Source of water	–0.160	–0.242	21	0.082	16	0.061 (X_4)	0.039 (X_7)	0.036 (X_{16})
X_{24}	Extent of weed infestation	–0.030	0.096	9	–0.126	21	0.171 (X_4)	0.083 (X_{16})	0.075 (X_7)

Residual effect = 0.486.

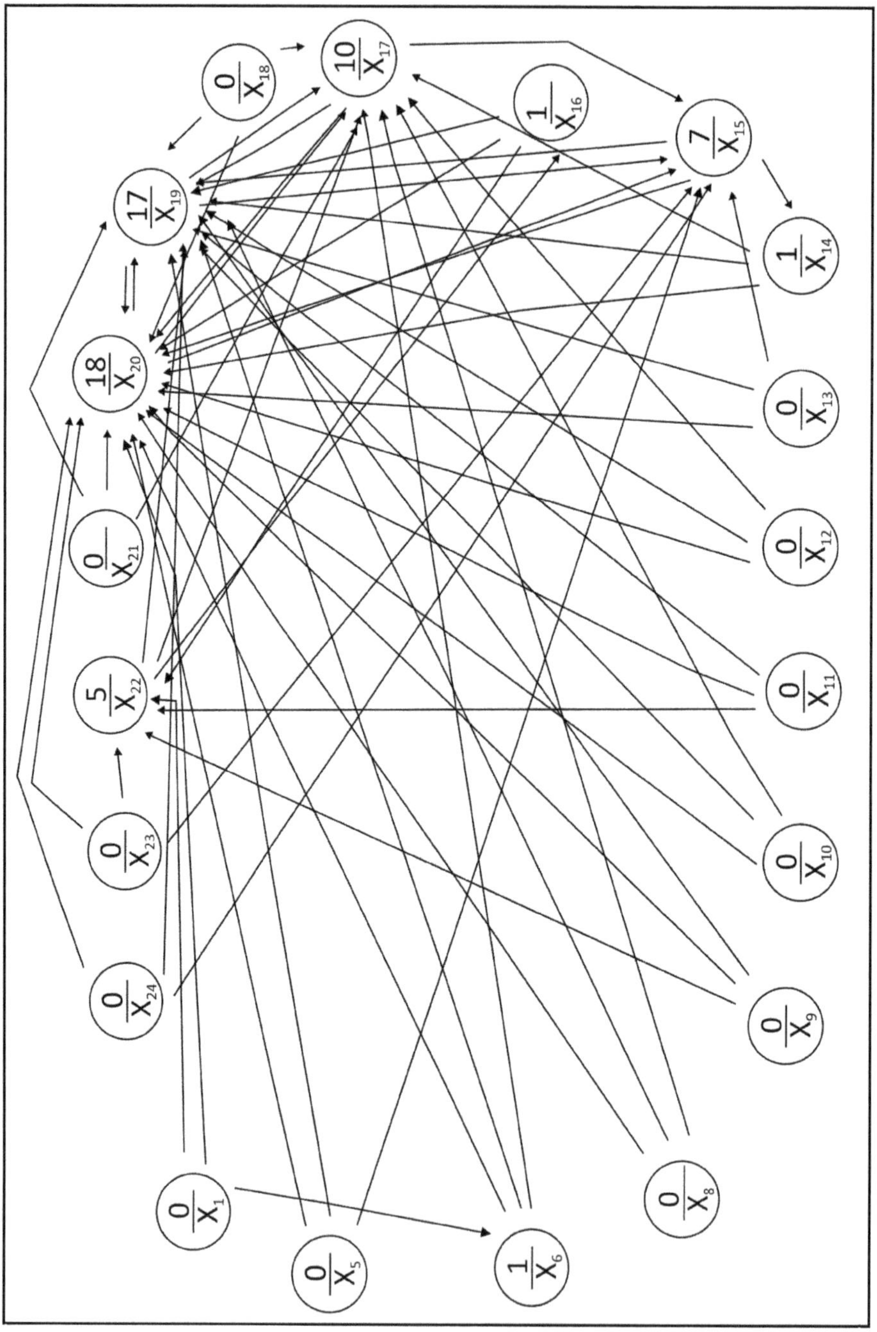

Figure 28: Diagrammatic Representation of the Variables through which Indirect Substantial Effects are Channeled (Adoption behaviour in North 24 Parganas).

extension agency contact (X_{13}) and knowledge (X_{20}), 6 pass through innovative proneness, 5 pass through age (X_1), 4 each pass through value orientation (X_{17}) and credit orientation (X_{16}), 3 pass through cosmopoliteness (X_{14}), 2 pass through education (X_5), one passes through caste (X_2). The residual effect is found to be 0.486 or in a way 48 per cent of the total variability, is left unexplained.

Dakshin Dinajpur

Direct Effects

A glance at Table 41 and Figure 29 reveals that out of 20 variables included in the path analysis, 11 variables have a positive direct effect and the remaining 9 variables have a negative direct effect on the adoption behaviour of the fish farmers towards scientific fish culture practices.

The ranking of variables based on their direct effect on the adoption behaviour of the fish farmers reveals that knowledge about scientific fish culture (X_{20}), extension agency contact (X_{13}), innovative proneness(X_{15}), extent of weed infestation (X_{24}), economic motivation (X_{19}) occupy the first five ranks in order whereas social participation (X_{10}), age (X_5), credit orientation (X_{16}), annual income (X_8), education (X_5) and mass media participation (X_{12}) obtain the last five ranks in the same successive order.

Indirect Effects

The total indirect effects channelized through other factors for each of the independent factors, are also found to be quite substantial. The ranking of these effects indicates that value orientation (X_{17}), risk orientation (X_{18}), cosmopoliteness (X_{14}), social participation (X_{10}), annual income (X_8) occupy the first five ranks with the largest total positive indirect effects on the adoption behaviour of the fish farmers towards the scientific fish culture, whereas land holding (X_9), extent of weed infestation (X_{24}), possession of fishing equipment (X_{11}), duration of water availability (X_{22}), and age (X_1) occupy the last five ranks exerting minimum indirect effects.

Substantial Indirect Effect

The first largest substantial positive effect on the adoption behaviour of the fish farmers is exerted by the variable economic motivation (0.289) through knowledge. The second largest substantial positive indirect effect of fish farmers is exerted by economic motivation (0.144) through extension agency contact. The third largest substantial positive indirect effect is exerted by cosmopoliteness (0.119) through extension agency contact. Out of 60 substantial indirect effects, 18 pass through the variable knowledge, nine pass through innovative proneness (X_{15}), eight pass through extension agency contact (X_{13}), seven passed through annual income (X_8), four each pass through mass media participation (X_{12}) and economic motivation (X_{19}), three each pass through education (X_5) and land holding (X_9), two pass through possession of fishing equipment (X_{11}), and one each through credit orientation (X_{16}) and extent of weed infestation (X_{24}). The residual effect is found to be 0.604, or in a way 60 per cent of the total variability is left unexplained.

Table 41: Path Coefficient Showing Direct, Indirect and Substantial Indirect Effects of Independent Variables on Adoption of Fish Farmers towards Scientific Fish Culture Practices in Dakshin Dinajpur

Code		*Coefficient*	*Effects*		*Effect*		*Substantial Indirect Effects are Channeled through*		
							I	II	III
X_1	Age	–0.230	–0.123	17	–0.107	20	0.067 (X_{20})	0.038 (X_8)	0.019 (X_9)
X_5	Education	0.140	–0.204	20	0.344	8	0.155 (X_{20})	0.081 (X_{15})	0.069 (X_{12})
X_6	Fish farming experiences	0.230	0.019	10	0.211	14	0.138 (X_{20})	0.078 (X_8)	0.072 (X_{12})
X_8	Annual income	0.240	–0.166	19	0.405	5	0.186 (X_{20})	0.084 (X_{24})	0.067 (X_{12})
X_9	Size of land holding	0.250	0.119	7	0.131	16	0.160 (X_{20})	0.084 (X_8)	0.061 (X_{12})
X_{10}	Social participation	0.320	–0.093	16	0.413	4	0.209 (X_{20})	0.084 (X_8)	0.078 (X_5)
X_{11}	Possession of fishing equipment	0.150	0.109	8	0.041	18	0.093 (X_{20})	0.055 (X_5)	0.052 (X_{13})
X_{12}	Mass media participation	0.490	0.153	6	0.337	9	0.235 (X_{20})	0.114 (X_{13})	0.084 (X_{15})
X_{13}	Extension agency contact	0.620	0.249	2	0.371	7	0.213 (X_{20})	0.0.091 (X_{19})	0.089 (X_{15})
X_{14}	Cosmopoliteness	0.480	0.061	9	0.419	3	0.249 (X_{20})	0.133 (X_{15})	0.119 (X_{13})
X_{15}	Innovative proneness	0.550	0.246	3	0.304	10	0.235 (X_{20})	0.090 (X_{13})	0.070 (X_{19})
X_{16}	Credit orientation	0.060	–0.165	18	0.225	13	0.093 (X_{20})	0.047 (X_{19})	0.028 (X_{11})
X_{17}	Value orientation	0.510	–0.072	13	0.582	1	0.262 (X_{20})	0.118 (X_{15})	0.114 (X_{13})
X_{18}	Risk orientation	0.450	–0.083	15	0.533	2	0.262 (X_{20})	0.111 (X_{15})	0.107 (X_{13})
X_{19}	Economic motivation	0.550	0.156	5	0.394	6	0.289 (X_{20})	0.144 (X_{13})	0.111 (X_{15})
X_{20}	Knowledge	0.740	0.444	1	0.296	11	0.130 (X_{15})	0.119 (X_{13})	0.102 (X_{19})
X_{21}	Size of water body	0.270	–0.006	12	0.276	12	0.169 (X_{20})	0.092 (X_8)	0.078 (X_9)
X_{22}	Duration of water availability	–0.010	0.003	11	–0.013	19	0.048 (X_9)	0.046 (X_{11})	0.41 (X_{16})
X_{23}	Source of water	0.080	–0.052	13	0.132	15	0.089 (X_{20})	0.069 (X_5)	0.056 (X_8)
X_{24}	Extent of weed infestation	0.280	0.233	4	0.047	17	0.120 (X_{20})	0.059 (X_8)	0.030 (X_{15})

Residual effect : 0.604.

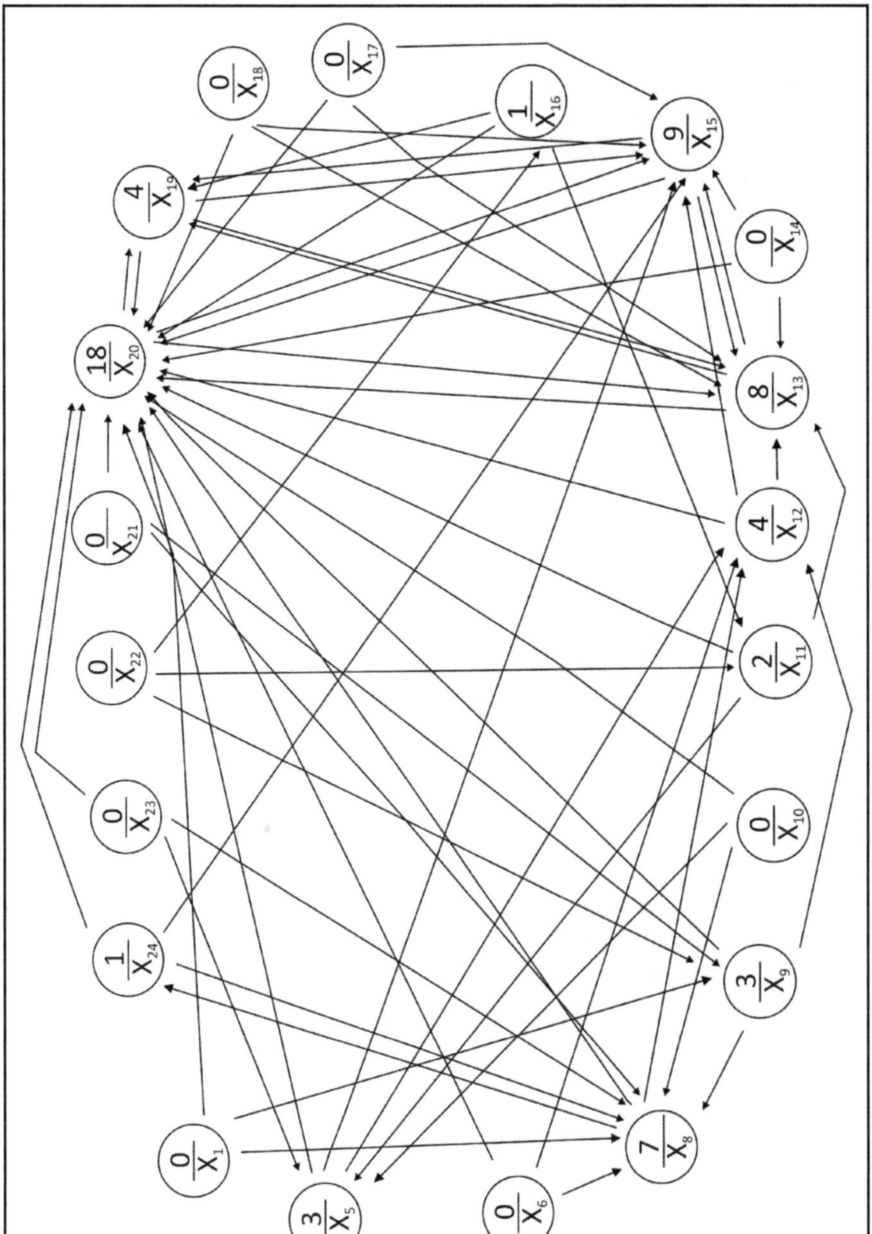

Figure 29: Diagrammatic Representation of the Variables through which Indirect Substantial Effects are Channeled (Adoption behaviour in Dakshin Dinajpur).

Hence from the above findings it can be concluded that in North 24 Parganas extension agency contact, innovative proneness, occupation, economic motivation, knowledge, cosmopoliteness and value orientation are found to be the dominant factors not only as factors having direct effects on the adoption behaviour of the fish farmers but also as factors through which most of the other factors influence the adoption level indirectly. Therefore, such variables as extension agency contact, innovative proneness, occupation, economic motivation, knowledge, cosmopoliteness and value orientation focus as dominant factors through which the adoption behaviour of farmers can be manipulated. These results are in strong conformity with the findings of stepwise multiple regression analysis. In Dakshin Dinajpur, knowledge, extension agency contact, innovative proneness, annual income, economic motivation, extent of weed infestation, value orientation, education and land holding are found to be the dominant factors not only as factors with direct effects on adoption behaviour of fish farmers but also as factors through which most of the other factors influence the adoption level indirectly. Therefore, such variables as knowledge, extension agency contact, innovative proneness, annual income, economic motivation, extent of weed infestation, value orientation, education and land holding focus as dominant factors through which the adoption behaviour of the farmers can be manipulated. These results are in strong conformity with the findings of stepwise multiple regression analysis.

PART-III

Descriptive Variable

Constraints to Adoption of Scientific Fish Culture

No study of the factors affecting the attitude and adoption behaviour of a group of fish farmers towards scientific fish culture is complete without duly taking into account the constraints and problems besetting the process of building and developing the desired attitude and adoption behaviour on the part of the fish farmers. Since the study aims at not only analyzing and presenting the status of the attitude and adoption behaviour as obtaining at the time of the study but also concerns itself with seeking and suggesting remedial measures, it is of considerable practical importance to formulate the constraints and problems in a meaningful and comprehensive manner rather than a slipshod manner. It is accordingly proposed to offer a brief historical account of how these problems and constraints were actually identified on the field through close interaction with the respondents.

As already mentioned, there are sixteen selected villages of two districts and out of these 16 villages, a sample of 240 fish farmers have been selected in conducting the study. In the beginning, when the research was still in the stage where attempts were made to build up a rapport with the local people of these sixteen villages, the following three general observations were recorded.

1. The fish farmers were facing some problems in regard to adopting scientific fish culture practices

Table 42: Summary Table of First Five Important Factors Showing Direct and Indirect Effects on the Basis of Path Analysis of the Respondents of North 24 Parganas and Dakshin Dinajpur

Sl.No.	Independent Variables	Direct Effect	Rank	Independent Variables	Indirect Effect	Rank
			Attitude			
North 24 Parganas						
1.	Knowledge (X_{20})	0.502	i	Value orientation (X_{17})	0.666	i
2.	Cosmopoliteness (X_{14})	0.467	ii	Social participation (X_{10})	0.473	ii
3.	Economic motivation (X_{19})	0.299	iii	Innovative proneness (X_{15})	0.418	iii
4.	Occupation(X_7)	0.291	iv	Risk orientation (X_{18})	0.414	iv
5.	Family type (X_4)	0.272	v	Extension agency contact (X_{13})	0.351	v
Dakshin Dinajpur						
1.	Economic motivation (X_{19})	0.426	i	Value orientation (X_{17})	0.688	i
2.	Knowledge (X_{20})	0.415	ii	Risk orientation (X_{18})	0.614	ii
3.	Innovative proneness (X_{15})	0.219	iii	Extension agency contact(X_{13})	0.489	iii
4.	Extent of weed infestation (X_{24})	0.161	iv	Cosmopoliteness (X_{14})	0.487	iv
5.	Cosmopoliteness (X_{14})	0.133	v	Mass media participation (X_{12})	0.453	v
			Adoption behaviour			
North 24 Parganas						
1.	Extension agency contact (X_{13})	0.411	i	Cosmopoliteness (X_{14})	0.584	i
2.	Innovative proneness (X_{15})	0.399	ii	Value orientation (X_{17})	0.562	ii
3.	Occupation (X_7)	0.359	iii	Risk orientation (X_{18})	0.373	iii
4.	Economic motivation(X_{19})	0.306	iv	Mass media participation (X_{12})	0.303	iv
5.	Knowledge (X_{20})	0.243	v	Size of water body (X_{21})	0.282	v
Dakshin Dinajpur						
1.	Knowledge(X_{20})	0.444	i	Value orientation (X_{17}),	0.582	i
2.	Extension agency contact (X_{13})	0.249	ii	Risk orientation (X_{18}),	0.533	ii
3.	Innovative proneness(X_{15}),	0.246	iii	Cosmopoliteness (X_{14}),	0.419	iii
4.	Extent of weed infestation (X_{24})	0.233	iv	Social participation (X_{10}),	0.413	iv
5.	Economic motivation (X_{19})	0.156	v	Annual income (X_8)	0.405	v

2. Because of their culture and traditions, they were having their own options for adopting scientific fish culture practices
3. The fishermen were found to be suggestive during the formal conversation with researcher, as far as improvement of their socio-economic condition through adoption of scientific fish culture practices

Keeping in mind the above mentioned observations, the following questions were incorporated in the interview schedule in order to get meaningful responses for the observation. An important objective of the present study is to identify the constraints and problems encountered by the fish farmers in their adoption of scientific fish culture. This kind of study helps to identify problems at various levels and of various nature like local, situational, technical and organizational which significantly affect the extent of the adoption of fish culture practices.

The Constraints Expressed by Fish Farmers of North 24 Parganas

The intensity of each problem is worked out by summing up all the individual scores obtained by the respondents on each of the problems. The problems are then categorized on their total constraint scores as the respondents perceived. The constraint ranked first is malicious poisoning of the fish pond. The second is poaching and theft of fish. Local non-cooperation of villagers rank third. High cost of manure and fertilizers rank fourth. High cost of artificial feed, non availability of high commercial feed, non-availability of fish seed in the village adequately and at proper time rank fifth, sixth and seventh respectively. Table 43 shows the problems and the ranking. The results of the study are presented below.

Malicious Poisoning of the Fish Pond

Problems of poisoning fish ponds are cited as a very serious problems by 51.67 per cent and serious problem by 35 per cent of the fish farmers.

Poaching and Theft of Fish

Majority of the fish farmers (62.5 per cent) report poaching and theft of fish as a very serious problem.

Local Non-cooperation of Villagers

Non-cooperation of villagers like objection to the application of manures, feeding, breaking the spill way, demanding free supply of fish, etc., are the problems faced locally by the fish farmers. This is expressed as a very serious problem by 45 per cent of the fish farmers.

High Cost of Manure and Fertilizers

High cost of manure and fertilizer is expressed as a very serious problem by 45 per cent and as a serious problem by 38.33 per cent of fish farmers.

High Cost of Artificial Feed

High cost of artificial feeds like mustard oil cake, groundnut oil cake, polished rice bran is reported as a very serious problem by 48.33 per cent respondents and as a serious problem by 30 per cent respondents.

Table 43: Constraints Encountered by the Fish Farmers in the Adoption of Scientific Fish Culture Practices in North 24 Parganas

Sl.No.	Problems	Not so Serious		Serious		Very Serious		Total Constraints Score	Rank
		Frequency	Per cent	Frequency	Per cent	Frequency	Per cent		
1.	Non-availability of qualitative fish seed timely and adequately in the village	43	35.83	42	35	35	29.17	112	7
2.	High cost of fish seed	77	64.17	31	25.83	12	10	55	18
3.	High cost of manure and fertilizer	20	16.67	46	38.33	54	45	154	4
4.	Lack of required soil and water testing facilities	70	58.34	34	28.33	16	13.33	66	15
5.	High cost of artificial feed	26	21.67	36	30	58	48.33	152	5
6.	Non availability of high commercial feed	42	35	32	26.67	46	38.33	124	6
7.	Lack of technical knowledge in judicious use of pesticides and other matters	84	70	23	19.17	13	10.83	49	20
8.	Malicious poisoning of the fish pond	16	13.33	42	35	62	51.67	166	1
9.	Poaching and theft of fish	32	26.67	13	10.83	75	62.5	163	2
10.	Local non cooperation of villagers	18	15	48	40	54	45	156	3
11.	Multi-ownership	92	76.67	19	15.83	9	7.5	37	22
12.	High involvement of middle men	43	35.83	59	49.17	18	15	95	9
13.	Lack of assured market	74	61.67	30	25	16	13.33	62	16
14.	Lack of credit facility	85	70.84	19	15.83	16	13.33	51	19
15.	High interest rate on loan	65	54.17	37	30.83	18	15	73	13
16.	High interest rate charge by money lender	59	49.17	42	35	19	15.83	80	11
17.	Procedural complications for obtaining loan from the credit organizations	40	33.33	54	45	26	21.67	106	8
18.	Inadequate transport facilities	92	76.67	15	12.5	13	10.83	41	21

Contd...

Table 43–Contd...

Sl.No.	Problems	Not so Serious		Serious		Very Serious		Total Constraints Score	Rank
		Frequency	Per cent	Frequency	Per cent	Frequency	Per cent		
19.	Harassment of police during transportation	104	86.67	12	10	4	3.33	20	23
20.	Inadequate follow up by extension personnel	75	62.5	35	29.17	10	8.33	55	18
21.	Lack of correct knowledge about the scientific fish culture	64	53.33	42	35	14	11.67	70	14
22.	Frequent flood	58	48.34	37	30.83	25	20.83	87	10
23.	Incidence of aquatic weeds, predators and weed fish	77	64.16	26	21.67	17	14.17	60	17
24.	High labour cost	63	52.5	36	30	21	17.5	78	12

Multiple responses were recorded.

Non Availability of High Commercial Feed

It is found that in adoption of scientific fish culture practices 38.33 per cent and 26.67 per cent fish farmers report non availability of high commercial feed as a very serious and as a serious problem respectively.

Non Availability of Fish Seed in the Village Adequately and at Proper Time

It is seen from table-43 that non-availability of qualitative fish seed in the village adequately and at proper time is expressed as a very serious problem and as a serious problem by 29.17 per cent and 35 per cent of the fish farmers respectively.

Procedural Complications for Obtaining Loan from the Credit Organizations

Procedural complications for obtaining loan from the credit organizations are reported as a very serious problem by 21.67 per cent and as a serious problem by 45 per cent of the fish farmers.

High Involvement of Middle Men

High involvement of middle men is cited as a very serious problem by 15 per cent and as a serious problem by 49.17 per cent of the fish farmers.

Frequent Flood

It is found that in the adoption of scientific fish culture practices 20.83 per cent and 30.83 per cent fish farmers report frequent flood as a very serious and as a serious problem respectively.

The other serious problems raised by the fish farmers include high interest rate charged by money lender, high labour cost, high interest rate on loan, lack of correct knowledge about various aspects of scientific fish culture, lack of required soil and water testing facilities, lack of assured market.

The constraints and problems as expressed by fish farmers of Dakshin Dinajpur

The assessment of the seriousness of each problem is carried out by summing up all the individual scores obtained by the respondents on each of the problems. The problems are then categorized on their total constraint scores as perceived by the respondents. The results are as follows. The constraint ranked first is poaching and theft of fish. The second is the high cost of manure and fertilizers. Non availability of fish seed in the village adequately and at proper time ranks third. Lack of required soil and water testing facilities ranks fourth. Malicious poisoning of the fish pond high cost of artificial feed, lack of credit facility are ranked fifth, sixth and seventh respectively. Table 43 shows the problems and the ranking.

Poaching and Theft of Fish

The majority of the fish farmers (60 per cent) reported poaching and theft of fish as a very serious problem.

High Cost of Manure and Fertilizers

High cost of manure and fertilizer is state to be a very serious problem by 63.3 per cent fish farmers.

Non Availability of Fish Seed in the Village Adequately and at Proper Time

It can be seen from the table that non-availability of qualitative fish seed in the village adequately and at proper time is termed as a very serious problem and as a serious problem by 54.17 per cent and 27.5 per cent of the fish farmers respectively.

Lack of Required Soil and Water Testing Facilities

It is found that in the adoption of scientific fish culture practices 47.5 and 31.7 per cent of the fish farmers report lack of soil and water testing facility as a very serious and as a serious problem, respectively.

Malicious Poisoning of the Fish Pond

The problem of poisoning fish ponds are cited as a very serious problems by 45.8 per cent and as a serious problem by 30.9 per cent of the fish farmers.

High Cost of Artificial Feed

The high cost of artificial feeds like mustard oil cake, groundnut oil cake, polished rice bran is reported as a very serious problem by 43.3 per cent respondents and as a serious problem by 37.2 respondents respectively.

Lack of Credit Facility

Financial constraints and lack of credit facilities to meet the expenditure on fish culture is reported as a very serious problem by 42.5 per cent of the fish farmers.

Inadequate Transport Facility

Inadequate transport facility is recorded as a very serious problem by 43.3 per cent of the fish farmers.

Local Non-cooperation of Villagers

Various form of non-cooperation of villagers such as objection for application of manures, feeding, breaking the spill way and demanding free supply of fish, etc., are the problems faced by the fish farmers. This is expressed as a very serious problem by 34.2 per cent of the fish farmers.

The other serious problems registered by the fish farmers include high cost of fish seed, lack of technical knowledge about judicious use of pesticides and other matters, lack of correct knowledge about various aspects of scientific fish culture, multi ownership of ponds and, non-availability of high commercial feed.

In the two districts the important constraints experienced by the respondents are non availability of fish seed in the village in adequate quality and on a timely basis, high cost of manure and fertilizers, high cost of artificial feed, non availability of high commercial feed, malicious poisoning of the fish pond, poaching and theft of fish, local non-cooperation of villagers, non availability of credit facility, procedural

Table 44: Constraints Encountered by the Fish Farmers in the Adoption of Scientific Fish Culture Practices in Dakshin Dinajpur

Sl.No.	Problems	Not so Serious		Serious		Very Serious		Total Constraints Score	Rank
		Frequency	Per cent	Frequency	Per cent	Frequency	Per cent		
1.	Non -availability of qualitative fish seed timely and adequately in the village	22	18.33	33	27.5	65	54.17	163	3
2.	High cost of fish seed	42	35	55	45.8	23	19.2	101	10
3.	High cost of manure and fertilizer	23	19.2	21	17.5	76	63.3	173	2
4.	Lack of required soil and water testing facilities	25	20.8	38	31.7	57	47.5	152	4
5.	High cost of artificial feed	27	22.5	41	34.2	52	43.3	145	6
6.	Non availability of high commercial feed	85	70.8	25	20.8	10	8.4	45	14
7.	Lack of technical knowledge in judicious use of pesticides and other matters	65	54.2	36	30.0	19	15.8	74	11
8.	Malicious poisoning of the fish pond	28	23.3	37	30.9	55	45.8	147	5
9.	Poaching and theft of fish	18	15	30	25	72	60	174	1
10.	Local non cooperation of villagers	49	40.8	30	25	41	34.2	112	9
11.	Multi-ownership	75	62.5	32	26.7	13	10.8	58	13
12.	High involvement of middle men	85	70.8	35	29.2	–	–	35	17
13.	Lack of assured market	89	74.2	24	2	7	5.8	38	16
14.	Lack of credit facility	29	24.2	40	3.3	51	42.5	142	7
15.	High interest rate on loan	95	79.2	22	18.3	3	2.5	28	18
16.	High interest rate charge by money lender	84	70.0	30	25	6	5	42	‘15
17.	Procedural complications for obtaining loan from the credit organizations	106	88.3	10	8.3	4	3.4	18	21
18.	Inadequate transport facilities	45	37.5	23	19.2	52	43.3	127	8

Contd...

Table 44–Contd...

Sl.No.	Problems	Not so Serious		Serious		Very Serious		Total Constraints Score	Rank
		Frequency	Per cent	Frequency	Per cent	Frequency	Per cent		
19.	Harassment of police during transportation	75	62.5	33	27.5	12	10	57	14
20.	Inadequate follow up by extension personnel	78	65	42	35	–	–	42	15
21.	Lack of correct knowledge about the scientific fish culture	71	59.2	35	29.1	14	11.7	63	12
22.	Frequent flood	101	84.2	12	10	7	5.8	26	19
23.	Incidence of aquatic weeds, predators and weed fish	100	83.3	15	12.5	5	4.2	25	20
24.	High labour cost	113	94.2	4	3.3	3	2.5	10	22

Multiple responses were recorded.

complications for obtaining loan from the credit organizations and inadequate transport facility, lack of necessary soil and water testing facilities.

Non-availability of qualitative fish seed at the right time is viewed as a limiting factor by the majority of the respondents of the two districts. As a result farmers can not stock fish seed after water storage in the ponds. This may affect the production level in the ponds. The present findings instead of the two districts are in line with findings of earlier studies by Bhaumik *et al.* (1988), Balasubramaniam (1988), Meeran and Jayaseelan (1996) and Alam (1997) and Talukdar (2000).

Organic manure in rural areas is usually diverted to agriculture and hence procurement becomes expensive for manuring fish ponds, as is seen in the two districts. This is in conformity with the findings of Ponappan (1982), Suresh *et al.* (1991) and Taufiq Rusdi (1994).

Soil and water testing facilities are essential for better production but are not adequately available in the two districts. This is in conformity with the findings of Meeran and Jayaseelan (1996).

High cost of artificial feed is an important constraint faced by the fish farmers of the two districts. Similar findings are reported by Suresh *et al.* (1990), Meeran and Jayaseelan (1996) and Bhaumik and Saha (1998).

Another constraint faced by the fish farmers of North 24 Parganas is that Non-availability of high commercial feed. Similar findings are reported by Mishra and Mishra (1999).

Some of the villagers of the two districts are found to be jealous of their fellow farmer's profit earned from fish culture (Bhaumik *et al.*, 1992 and Bhaumik and Saha, 1998).

One of the important social problems experienced by the fish farmers of the two districts is poaching and theft of fish from the ponds. Many farmers find that in spite of watch and ward, night poaching can not be averted owing to the small size of the water area of the ponds, which are easily poachable. Further, this problem gets aggravated if a farmer is not supported by village leaders or is unable to manage problems tactfully. The present finding is in conformity with the findings reported by Bhaumik *et al.* (1992). However, Subhas Chandra (1986), Balasubramaniam (1988), Reddy (1997) and Talukdar (2000) report that poaching of fish is not a serious problem.

It is also reported by the respondents of the two districts that villagers and farmers object to the application of manures like cow dung, poultry manure, etc., and recommended feeds like mustard oil cake and rice bran, etc. It is reported by the fish farmers that villagers feel that the use of pond by the village community at large should not be curbed by way of being restricted for the benefit of a single person even if he or she owns it. The present findings are in conformity with the findings reported by Bhaumik *et al.* (1992) and Le Xuan Sinh (1996).

Nine important suggestions have been given by the fish farmers for adoption of scientific fish culture as indicated in Table 45.

Table 45: Suggestions of Fish Farmers for Adoption of Scientific Fish Culture

Suggestions	*North 24 Parganas*		*Dakshin Dinajpur*	
	Frequency	*percentage*	*Frequency*	*percentage*
The establishment of fish hatchery at the Gram Panchayat level	74	61.67	78	65
Minimization of interference by the middle men	77	64.17	54	45
The establishment of fish feed mill	82	68.33	51	42.5
The establishment of a soil and water testing centre	83	69.17	69	57.5
The formation of fishermen's cooperative society, farmers' club, etc.	71	59.17	65	54.17
Need of bank loan through easy procedure	64	53.33	54	45
Need of arrangements for imparting training about improved fish culture	75	62.5	63	52.5
Providing easy transport of fish.	53	44.17	62	51.67
The establishment of a Govt. regulated fish market	78	65	72	60

Out of these nine suggestions, the most important suggestions as given by all the fish farmers of North 24 Parganas include the establishment of a soil and water testing centre, the establishment of fish feed mill, the establishment of a govt. regulated fish market, minimisation of interference by the middle men, need of arrangements for imparting training about improved fish culture, the establishment of fish hatchery at the Gram Panchayat level, the formation of fishermen's cooperative society, farmers' club, etc. In Dakshin Dinajpur the most important suggestions given by most of the fish farmers include the establishment of fish hatchery at the Gram Panchayat level, the establishment of a govt. regulated fish market, establishment of a soil and water testing centre, the formation of a fishermen's cooperative society, farmers' club, etc., need of arrangements for imparting training about improved fish culture, providing easy transport of fish.

Chapter 6

Summary and Conclusion

Fish has occupied an important place in the global food basket as a safe and cheap source of animal protein with high consumer acceptability. India with a vast coastal and inland waterway resources and an ever wide gap between the optimum per capita protein intake and actual in the context of second largest population in the world, represents a land that badly needs to improve in all possible ways in the exploration of its immense potential for pisciculture and allied activities. Despite gaps here and there, it may broadly be recognized that India has fairly succeeded in meeting this crying need. India has already emerged as one of the global leaders in the fisheries and aquaculture sector through development and adoption of a host of technologies for increasing fish production. The fish production of the country has increased from 0.75 million tonnes in 1950 to 9.72 million tonnes at present, a more than twelve fold increase (Ministry of Agriculture, 2015). At the same time the share of inland fisheries has gone up from 29 percent to over 60 percent (Ministry of Agriculture, 2015). India's performance is, however, to be assessed not merely in terms of absolute fish production but rather also in the light of the success in harnessing the available resources and meeting the need to provide people with adequate fish protein. In this view of the matter the picture that emerges is of mixed success. While the fish production has steadily been on the increase, there yet remains a leeway to go and the per capita consumption of fish still lags behind the desired level. The per capita availability of fish in India is about 8 kg while the world average is 18 kg as against the recommendation of 11 kg by World Health Organization (WHO). India exported fish products worth about Rs. 30213 crores during 2013-14. The fisheries sector contributed 4.75 percent share in agriculture Gross Demostic Product (GDP) and 1.1 percent share in the total GDP. Annual production potential for marine and inland fish production has been estimated to be 3.9 million tonnes and 4.5 million tonnes, respectively. Recent trends in fish production indicated that marine fish production has almost reached the plateau and there is not much scope for increasing further the marine fish production, but aquaculture has been one of the fastest growing agricultural sector of the country.

The average growth rate of over 6 percent per year in the last two decades is testimony to the potential in the fishery sector. The sector is endowed with present fish water aquaculture production of about 2.4 tonnes/hectare, which contributes about 95 percent of the total aquaculture production and accounts for about 40 percent of the total fish production of the country, has been possible through phenomenal growth in development of carp farming in small inland water bodies, that is, ponds. Thus, aquaculture is expected to play a pivotal role in the stepping up of fish production in the immediate future.

In Chapter 1 an attempt has been made to introduce the problems of the present research on a broad basis: statement of the problem, objectives of the study, scope of the study, limitations of the study and layout of the study.

West Bengal has been playing a significant role in regard to fish culture since time immemorial. West Bengal possesses vast inland aquatic resources. The total fish production of the state 14.90 lakh tonnes with 1.52 lakh tonnes from marine sector and 13.38 lakh tonnes from inland resources (Hand book of Fisheries Statistics, 2012-13). The contribution of the state to the total fish production of India during 2004-05 is about 15.32 percent. With its age old importance of fish culture as a means of livelihood, as a major economic activity and finally as a way of life with a large section of the fisherman population, West Bengal offers a veritable case for the conduct of a study of how a qualitative change may be initiated in the participation of the fish farmers themselves in the process by way improving their motivation towards adoption of scientific fish culture. This will of course form part of large process of bringing about the desired transformation of the traditional activities of fish culture to bring them into line with scientific management of available resources by application of modern technologies and practices.

Keeping in view this background the present study entitled "Factors affecting attitude and adoption behaviour of fish farmers towards scientific fish culture in two districts of West Bengal, India" was taken up with the following specific objectives:

- ☆ To study the selected socio-economic, psychological and communication attributes of the fish farmers in the study area
- ☆ To ascertain the attitude and adoption behaviour of fish farmers towards scientific fish culture
- ☆ To determine the factors affecting the attitude and adoption behaviour of fish farmers with regard to scientific fish culture
- ☆ To bring out the relationships of the socio-economic, psychological and communication characteristics of the fish farmers with their attitude and adoption of scientific fish culture
- ☆ To find out the constraints in the adoption of scientific fish culture as experienced by the fish farmers

In Chapter 2, the researcher has attempted to present the review of past works in connection with present study in a systematic manner. This chapter also include conceptual orientation, an attempt has been made to convey an understanding of the technical terms and concepts, etc., related to the study in their theoretical perspective.

The relevant literature was reviewed on the following aspects of the study

- ☆ Conceptual orientation
- ☆ Profile of fish farmers
- ☆ Attitude towards agricultural and allied farm enterprises
- ☆ Association of attitude with independent variables
- ☆ Adoption of recommended practices of agricultural and allied farm activities
- ☆ Association between adoption of recommended practices with independent variables
- ☆ Constraints to adoption of scientific fish culture practices

Chapter 3 entitled as research setting deals with background information of the area and the people. No interpretation can be drawn unless a clear understanding of the characteristics of the area and attitude of the people towards scientific technology are known. The study was carried out in two districts namely, North 24-Parganas and Dakshin Dinajpur with total annual fish production 1.82 lakh ton and 0.54 lakh ton, respectively. The information with maps of country, state, districts and blocks have also been placed.

In Chapter 4, the methodology adopted in the study has been discussed. Two districts namely, North 24 Parganas and Dakshin Dinajpur district of West Bengal, India were selected purposively in view of the vast and diverse inland fishery resources ideally suited for taking up scientific fish culture. Among twenty two Development Blocks in North 24-Parganas four blocks namely Amdanga, Hasnabad, Habra-I and Habra-II and from eight Development Blocks in Dakshin Dinajpur district four blocks namely Balurghat, Kumarganj, Tapan and Gangarampur were selected. From each block, two villages were selected with the help of a random number table following the method of simple random sampling without replacement technique. In total eight villages of each district served as the representing unit for this study. The selected villages of each district which formed the universe were stratified on the basis of the concentration of the fish farmers. A stratified random sample with proportional allocation from each stratum was drawn. Total sample size is 240, of which 120 for each district.

In the present study, a scale developed by Nagarajaiah (2002) was used to measure the attitude of fish farmers towards scientific fish culture and the adoption behaviour of fish farmers towards scientific fish culture was measured using the modified scoring procedure followed by Sinha and Kolte (1974). The scale developed by the earlier researcher with suitable modifications and schedules developed were used for measuring the variables. Total independent variables included in the study were 24. Further, the constraints faced by the fish farmers in the adoption of recommended practices were also studied.

A pre-tested interview schedule specially developed for the purpose of the study was administered to collect primary data through personal interview method from the respondents. Apart from primary data, use was also made of secondary data like government publications, research papers, reports and other published documents.

The collected data were scored, tabulated and analysed. Keeping in view the objectives of the present study, different statistical tools *viz.*, frequency and percentage, correlation test, multiple regression analysis and path analysis were employed in the analysis and interpretation of data.

In chapter 5, the findings and discussion of the present study have been presented and discussed as under.

Socio-Personal, Socio-Economic, Communication, Psychological and Situational Characteristics of Fish Farmers

The following traits are observed on the whole in the case of the majority of fish farmers in North 24 Parganas: a) they belonged to the middle aged group. b) overall education level was up to medium school level that is, primary to middle high school. c) they belonged to medium social participation, low land holding and high income group as they have other source of income. Whereas in Dakshin Dinajpur the majority of fish farmers : a) they belonged to the middle aged group. b) overall education level was up to medium school level that is, primary to middle high school. c) they belonged to low social participation, small to medium land holding and high income group as they have other source of income.

Similarly, in regard to a set of psychological and communication variables, most of the farmers of North 24 Parganas evinced the following characteristics: a) they belonged to high to medium level of innovative proneness and medium level of value orientation, risk orientation and economic motivation. b) they belonged to low to medium level of credit orientation, and possessed medium to high level of knowledge. c) they exhibited medium level of mass media participation and medium to low level of extension agency contact while their cosmopoliteness was high. Most of the farmers of Dakshin Dinajpur evinced the following characteristics: a) they belonged to medium level of risk orientation, economic motivation and high to medium level of value orientation and possessed either high or low level of innovative proneness b) they belonged to low level of credit orientation, and possessed medium level of knowledge. c) they exhibited low level of mass media participation and medium to low level of extension agency contact while their cosmopoliteness was low.

In North 24 Parganas, majority of the ponds were medium to small sized and water holding capacity were low to medium and rainfed. Most of the ponds were infested with weeds and the extent of weed infestation was low. On the other hand Dakshin Dinajpur, majority of ponds were small sized, medium to low in water holding capacity and rainfed. Most of the ponds were infested with weeds and the extent of weed infestation was moderate.

In North 24 Parganas, most of the fish farmers (38.3 per cent) possessed medium level of knowledge on scientific fish culture. Majority of the fish farmers were aware of simple and routine practices but they did not possess the specific scientific knowledge regarding nutrients required for production of fish food organism, name of fish grows well in weed infested ponds, advantages of using inorganic fertilizers, names of piscicides, application of mohua oil cake, harvesting period, minimum depth of water required for fish culture, time gap required between manure application and stocking of fish seed, correct feeding rate of supplementary feed and the necessity of checking growth. This showed that knowledge on scientific fish culture has not fully percolated to the fish farmers' level and the knowledge so acquired was partial. On the other hand in Dakshin Dinajpur a large section of the fish farmers (43.3 per cent) possessed medium level of knowledge on scientific fish culture. Majority of fish farmers were aware of simple and routine practices but they did not have the specific scientific knowledge regarding minimum depth of water, nutrients required for production of fish food organism, advantages of using inorganic fertilizers, names of piscicides, application of mohua oil cake, method of eradication of predatory and weed fishes, the optimum size of harvesting, harvesting period, the necessity of checking growth. This showed that knowledge on scientific fish culture has not fully percolated to the fish farmers' level and the knowledge so acquired was partial.

The attitude of the majority of the fish farmers in North 24 Parganas varied from medium favourable to more favourable attitude towards scientific fish culture practices. However, nearly 1/5th of them were in less favourable category. As against this in Dakshin Dinajpur, the said attitude of the majority of fish farmers varied from medium favourable to more favourable. However, nearly ¼th of them were in less favourable category.

The adoption score on scientific fish culture practices as obtained by the fish farmers was found to vary in the two districts in the following manner. In North 24 Parganas, the majority of the farmers have obtained medium adoption score. The mean adoption score (76.57) obtained by the fish farmers also revealed the medium level of adoption. On the other hand in Dakshin Dinajpur, the majority of the farmers (nearly ¾th) belonged to medium to low adoption category. The mean adoption score (67.51) obtained by the fish farmers also revealed the medium level of adoption.

Relational Analysis

In North 24 Parganas, out of 24 variables included in the study, 10 variables were positively and significantly related with the attitude of fish farmers towards scientific fish culture practices. Among the variables, all the psychological (except credit orientation), communication factors (except extension agency contact) education, annual income, social participation were positively and significantly related with attitude towards scientific fish culture. But in Dakshin Dinajpur, among the variables, all the communication, psychological variables (except credit orientation) and situational variables (except duration of water availability) and education, fish farming experience, annual income, social participation, land holding were positively and significantly related with attitude of fish farmers towards scientific fish culture and only one variable that is, family type was negatively and significantly correlated with attitude of fish farmers towards scientific fish culture practices.

Out of 24 variables included in the study in North 24 Parganas, 12 variables were positively and significantly related with adoption level of fish farmers and only two variables (age and caste) was negatively and significantly related with adoption level of fish farmers. All the psychological variables (except credit orientation), communication variables, occupation, land holding (socio- economic) and size of water body and duration of water availability (situational variable) had positive significant correlation with adoption. Most of the socio-personal, socio-economic and situational variables could not establish any significant relationship. On the other hand in Dakshin Dinajpur, out of 24 variables included in the study, 14 variables were positively and significantly related with adoption level of fish farmers. All the psychological (except credit orientation), communication, socio-economic (except occupation and possession of fishing equipment) variable, fish farming experience, size of water body and extent of weed infestation had positive significant correlation with adoption. Most of the socio-personal and situational variables could not establish any significant relationship.

In North 24 Parganas, all the 24 variables, when pooled together explained 56 percent variation in the attitude of the fish farmers and significant contribution to the variation was attributed to variables *viz.*, caste, family size, annual income, cosmopoliteness, innovative proneness, knowledge and size of the water body. Whereas in Dakshin Dinajpur, all the twenty four variables, when pooled together explained 27 percent variation in the attitude of the fish farmers and significant contribution to the variation was attributed to variables *viz.*, education, annual income and social participation.

In North 24 Parganas, all the twenty four variables when taken together explained 71 percent variation in the adoption behaviour of the fish farmers. The ten variables *viz.*, age, family type, occupation, land holding, social participation, extension agency contact, innovative proneness, credit orientation, knowledge and source of water contributed significantly for the variation in the adoption behaviour of the fish farmers. On the other hand in Dakshin Dinajpur, all the twenty four variables when taken together explained 73 percent variation in the adoption behaviour of fish farmers. The eight variables *viz.*, age, education, annual income, land holding, extension agency contact, innovative proneness, knowledge and extent of weed infestation contributed significantly for the variation in the adoption behaviour of fish farmers.

In North 24 Parganas, path analysis revealed that knowledge about scientific fish culture exhibited highest positive direct effect on the attitude of fish farmers towards scientific fish culture followed by the variables, cosmopoliteness, economic motivation, occupation and family type. Among the total indirect effects, the highest and lowest favourable contributions were by the variables value orientation, age respectively. Majority of the substantial indirect effects were channelised through value orientation, knowledge, cosmopoliteness, economic motivation, family type. Hence, these factors emerged as most dominant variables in influencing the attitude of fish farmers whereas in Dakshin Dinajpur, path analysis revealed that economic motivation exhibited highest positive direct effect on the attitude of fish farmers towards scientific fish culture followed by the variables knowledge about scientific fish culture, innovative proneness, extent of weed infestation and cosmopoliteness. Among the total indirect effects, the highest and lowest favourable contributions were by the variables value orientation and age, respectively. Majority of the substantial indirect effects were channelised through knowledge, economic motivation, value orientation, innovative proneness and duration of water availability. Hence, these factors emerged as most dominant variables in influencing the attitude of fish farmers.

In North 24 Parganas, path analysis revealed that extension agency contact exhibited highest positive direct effect on adoption behaviour of fish farmers followed by innovative proneness, occupation, economic motivation, knowledge about scientific fish culture. Among the total indirect effects, the highest and lowest contributions were by the variables cosmopoliteness and caste respectively. Majority of the substantial indirect effects were channelized through economic motivation, family type, occupation, extension agency contact, knowledge and innovative proneness. But in Dakshin Dinajpur, path analysis revealed that knowledge about scientific fish culture exhibited highest positive direct effect on adoption behaviour of fish farmers followed by extension agency contact, innovative proneness, extent of weed infestation and economic motivation. Among the total indirect effects, the highest and lowest contributions were by the variables value orientation and age respectively. Majority of the substantial indirect effects were channelized through knowledge, innovative proneness, extension agency contact, annual income, mass media participation and economic motivation, education and land holding.

Constraints to Adoption of Scientific Fish Culture

Major constraints as expressed by the fish farmers in the adoption of scientific fish culture practices in North 24 Parganas included the non availability of fish seed at proper time, high cost of manure and fertilizers, lack of required soil and water testing facilities, high cost of artificial feed, non availability of high commercial feed, malicious poisoning of the fish pond, poaching and theft of fish, local non-cooperation of villagers, high involvement of middle men, procedural complications for obtaining loan from the credit organizations, inadequate transport facilities. Whrereas in Dakshin Dinajpur non availability of fish seed in the village adequately and at proper time, high cost of manure and fertilizers, lack of required soil and water testing facilities, high cost of artificial feed, malicious poisoning of the fish pond, poaching and theft of fish, local non-cooperation of villagers, non availability of credit facilities, inadequate transport facilities were the major constraints expressed by the fish farmers in adoption of scientific fish culture practice.

Chapter 6 outlines the summary and conclusion based on the study. Attempts have also been made to draw some implications and recommendations on the basis of results obtained in the study. Finally on the basis of the implications some areas have been identified for future research in the domain.

Implications and Recommendations

The findings of the present investigation have brought to focus the following implications for consideration by appropriate authorities and agencies for developing suitable strategies and measures aimed at increasing fish productions –

1. According to the study there is adequate scope for bringing about improvements in the knowledge level and also for developing a more favorable attitude of fish farmers towards scientific fish culture. For this purpose it is necessary to strengthen and intensify the related extension efforts.
2. Keeping in view the need to improve the fishermen's- both the policy makers and extension functionaries should make adequate use of various teaching methods effectively like demonstration, field day, On-farm testing, exhibition, film show, educational tour, campaigns, farm clinic, seminar, workshop and information communication technology like radio, TV, different audio visual aids and internet, etc. for raising the level of knowledge on different aspects of scientific fish culture together with marketing and cultural practices.
3. This process of utilizing various teaching methods should also be supplemented by appropriate measures, conceived at the level of the policy makers and extension specialists, which are aimed at suitably manipulating such crucial variables as knowledge, fish farming experience, economic motivation, education and land holding so as to bring about the desired changes in the attitude and adoption behaviour of the fishermen towards scientific fish culture.
4. Extension activities should focus on popularizing those scientific practices which are needed to be incorporated such as; testing of soil and water, use of inorganic fertilizers, application of lime, initial manuring and organic manures, use of supplementary feeds, feeding rate, stocking density and ratio of stocking.
5. With a view to brining about new changes in the fish farmer's adoption and attitude, comprehensive and suitable extension programme may be arranged such as organising study tours, field trips, visit to result demonstration and front line demonstration units etc, for fish farmers to observe the achievements of other farmers with their own eyes.
6. In order to improve the process of reorienting the fishery extension system and to provide technical and input support to the farmers to enhance knowledge and adoption level, the authorities should arrange to formulate and monitor visit schedule of extension officials along with introduction of a system of evaluation at apex level.
7. Fishermen should be encouraged to opt for insurance coverage for fish crops to guard against risk involved in the adoption of scientific fish culture.
8. There is need at the level of the educational, research and extension institutions and organizations to update the extension syllabus suitably

to incorporate the results of this study to the extent possible in so far as they relate to vital aspects of fishery extension methodologies of analysis of emerging scenario in fishery and its characterization and likely future perspectives, and appropriate methodologies and techniques for modulation and encapsulation aiming at increased professionalism as extension practitioner for equipping the farmers to cope up with the changing environment. Extension professionals need to integrate all aspects of technology generation–refinement and dissemination. For integration of knowledge, talented experts are required and extension professionals need to possess high degree of skills in analysis, diagnostics, communication and motivation. It is high time that extension professionals acquire the skill to work in teams cutting across disciplines and breaking boundaries to help the farmers in integrated manner.

9. Although no extension system can be expected to offer a unified model to resolve all the problems of a complex agricultural system, endeavours may be made to bring about suitable changes in the existing systems with a view to making them vibrant enough to respond the crying needs of the personnel involved as revealed in this study. In particular, the system may be sought to be made more informed by farming situation specific approaches that take care of farmers' participation and other related matters. The extension strategy may also be suitably reorganized so as to reorient it towards creating amongst the farmers a potent inclination to conduct his farming operations in such a way as to make them more responsive to the needs of becoming competitive in the market economy by way of rendering the operations more market centred through various devices such as product diversification, fish processing and value addition for fish products and exploration of new areas in fish production and the like.

Suggesionst for Future Research

1. The present study was limited to only two districts of West Bengal. Therefore, it is suggested that further investigations may be taken up in rest of all districts of West Bengal. Such broadening the locale of the study facilitate generalizations regarding attitude and adoption behaviour of fish farmers which would have immense significance in making future extension strategies and programme planning for promotion of scientific fish culture at the state.
2. At present there are no proper scales for measuring the knowledge and attitude of extension personnel working with fish farmers. Hence, the required scales may be developed for measuring the knowledge and attitude of fisheries extension personnel. This may help to explore the possibilities of improving the skills and capabilities of the change agents to serve their clientele better.
3. In the present study the independent variables have been found to explain 56 and 71 per cent of the variation in the attitude and the adoption

behaviour of fish farmers towards scientific fish culture respectively in North 24 Parganas and the independent variables have been found to explain 27 and 73 per cent of the variation in attitude and adoption behaviour of fish farmers towards scientific fish culture respectively in Dakshin Dinajpur. Therefore, studies on additional independent variables may be initiated to supplement the findings of present study.

4. Detailed investigation on the pattern and effectiveness of the various existing utilization of extension methods and communication channels by fish farmers needs due consideration.
5. Future researches may do well to undertake empirical study on Field level constraints of scientific fish culture in varying conditions prevailing in different districts of West Bengal.

References

Alam, S.S. 1997. Economic evaluation study of FFDA programme for freshwater development. Sonitpur district, Assam, D.F.Sc. Dissertation, CIFE, Mumbai.

Alikunhi, K.H. 1957. Fish culture in India. *Fm. Bull. Indian Coun. Agri. Res.,* **20** : 14 pp.

Allport, G. 1935. Attitudes. In : A hand book of Social Psychology. Edited by Murichson, G. Lorcestor, Clark University Press.

Amid, I. Ismail. 2005. Reliability and Validity of Brief Measures of Oral Health-Related Knowledge, Fatalism and Self-Efficiency in Mothers of African, American Children. *Pediatr Dent*. 27(5): 122-428.

Anonymous, 1976. First 138 case studies of composite fish culture in India. *Bull. Cent. Inland Fish Res. Inst.* **23** (mimeo), CIFRI, Barrackpore.

Anonymous, 1984. Seventh workshop on All India Coordinated Research Project on composite fish culture and fish seed production held at FARTC, CIFRI, Kaushalyaganga, Orissa, pp. 123.

Anonymous, 1992. Ministry of Rural Development Circular No. 280112/10191-IRD III (Vol. II) dt. 15th Nov., 1991. *Gramin Vikas Newsletter,* **8** (1) : 14-20.

Anonymous, 2004-05. Annual Report, Department of Fisheries, Aquaculture, Aquatic resources and Fishing Harbours, Govt. of West Bengal.

Anonymous, 2006. District Statistical Hand Book-Dakshin Dinajpur. Bureau of Applied Economics and Statistics, Govt. of W.B., 2006.

Anonymous, 2011. Report of the working group on Development and management of Fisheries and Aquaculture. For the XI five year plan by Planning Commission, Govt. of India

Anonymous, 2013. Hand Book of Fisheries Statistics, Dept. of Fisheries, Govt. of W.B.

Anonymous, 2014. Hand book of Fisheries Statistics, 2014, Dept. of Animal Husbandary, Dairying and Fisheries, Ministry of Agriculture, Govt. of India

Arivand, N. 1999. Homestead Agro-forestry system in Kerala – Farmers perception and attitude, M.Sc. thesis (Unpub.), U.A.S., Bangalore.

Atkinson, R.L., Atkinson, R.C. and Hilgard, E.R. 1983. Introduction to psychology, pp. 543-550, 8th Ed., Chapter 17, Harcourt Brace Jovanovich, Inc, New York.

Awasthi, H.K., Singh, P.R. and Sharma, R.N. 2000. Knowledge and attitude of dairy farmers towards improved dairy practices. *Maharashtra J. Exten. Educ.* **19** : 290-292.

Ayyappan, S. 2008. Fish for diversified agriculture. Proceedings of Third National Conference on KVK.27-29 December, 2008, Div. of Agril. Extension, ICAR.

Ayyappan, S. and Jena, J.K. 2003. Indian aquaculturist and aquaculture : Present status and future project. Fish for all : National Lunch held at Kolkata, India, December 18-19, 2003 organized by ICLARM, Govt. of India and W.B. and M.S. Swaminathan Foundation, pp. 12-14.

Ayyappan, S. and Jena, J.K. 2006. On a fast tract of development. Survey of Indian Agriculture 2006. The Hindu.

Ayyappan, S. and Venkateshwarlu, G. 2002. Production, much below potential. *Survey of Indian Agriculture*, 2002. The Hindu.

Balasubramaniam, S. 1988. Analysis of technology transfer effectiveness in inland fish farming. Ph.D. thesis (unpub.), TNAU, Coimbatore.

Balasubramaniam, S. and Kaul, P.N. 1984. Adoption of improved practices by Fish Trawler owners in Kerala. *Indian J. Extn. Edu.*, **20** (3 and 4) : 35-42 pp.

Balasubramaniam, S. and Kaul, P.N. 1985. Adoption of improved practices by traditional fishermen in Kerala. *Indian J. Extn. Edu.*, **21** (3 and 4) : 80-88.

Balasubramaniam, S., Bihari, B and Mohan, B. 2001. Socio-economic status of marine fishermen in two fishing villages of Orissa. *Fishery Technology*, **38** (1) : 51-55.

Ban, A.W.V. and Hawkins, H.S. 1988. Agricultural Extension. Longman group, UK, Limited. Longman House, Burmi, M.U. Hariow, England.

Bandyopadhyay, M.K. and Pradeep K.K. 2002. Socio-economic traits and fisheries enhancement in Oxbow Lakes of North Bengal. Paper presented at National Symposium on "Fisheries Enhancement in Inland waters – Challenges ahead- 27-28 April, 2002, CICFRI, Barrackpore, West Bengal.

Basavakumar, K.V. and Yaligar, D.M. 2000. Impact of training on the knowledge and opinion of trainees on brackish water prawn farming. *Karnataka J. Agril. Sci.* **13** (3) : 707-711.

Beal, G.M. 1958. Information sources in decision making process. The Research Clinic on Decision Making, Washington, Pullman, State College : 136-151.

Beal, G.M. and Sibley, D.N. 1967. Adoption of agricultural technology by the Indians of Guatemala, Department of Sociology and Anthropology, Ames. Iowa State University, *Rural Sociological Report.* 62.

Beal, G.M., Rogers, E.M. and Bohlen, J.M. 1956. Validity of the concept of stages in the adoption process. Ames. Iowa Agricultural Experiment Station, Journal Paper J-3100.

Beal, G M and Bholen, J. 1957. The diffusion process, Iowa Agriculture Experiment Station, Special Report No. 18.

Beerannavar, B. 1995. A study on knowledge and adoption of improved dairy management practices by farmers trained by KVK, Hanumanamatti, M.Sc. (Ag.) thesis (Unpub.), U.A.S., Dharwad.

Bhakhri, V. 1985. Impact of recommended dryland agricultural technology on attitude knowledge and adoption of farmers in Jammu district (J and K). M.Sc. (Ag.) thesis (unpub.), APAU, Hyderabad.

Bhaumik, U., Pandit, P.K., Chatterjee, J.G. and Halder, D.D. 1988. The problems of exotic carp culture and suggested measures – The fish farmers perception. *J. Inland Fish Soc. India.* **20** (2) : 38-42.

Bhaumik, U., and Saha, S.K. 1998. Need for modification of composite fish culture technology in West Bengal as perceived by the fish farmers. Current and Emerging trends in aquaculture. Daya Publishing House, New Delhi, pp. 348-354.

Bhaumik, U., Pandit, P.K. and Karmakar, H.C. 1992. Adoption behaviour of fish farmers towards composite fish culture. *J. Inland Fish. Soc. India.* **24** (1) : 50-55.

Bhimachar, B.S. 1942. Report on a survey of the fisheries of the Mysore State. *Fish. Bull., Mysore,* **(1)** : 39p.

Biswas, M. 1990. A study on the knowledge and adoption of rice production technology by the farmers in Andaman district. M.Sc. (Ag.) thesis (Unpub.), Andhra Pradesh Agricultural University (APAU), Hyderabad.

Biswas, A., Acharjee, S.K. and Haque, M.A. 1991. Adoption of composite fish culture in the context of some psychological orientation. *Environment and Ecology.* Vol. **9** (3) : 661-663.

Bohlen, J.M. and Beal, G.M. 1960. Sociological and social psychological factors related to credit used patterns. Unpublished paper presented at Annual Conference of TVA Co-operative, Knoxville, Tenesse, Ames, Mimeographed. Department of Economics and Sociology, Iowa State University of Science and Technology.

Bose, A.B. 1965. Communication for agricultural innovation. Agric. Res. **5** : 324.

Bose, J.C. 1989. A study on consequences in adoption of fish culture practices. M.Sc. thesis, TNAU, Coimbatore.

Bose, S.P. 1995. Socio-cultural factors in farm efficiency, *Indian J. Extn. Edu.* **1** : 194.

Braj, M. Balasubramaniam, S., Kandoran, M.K. and Thomas, M. 1993. Technological changes among the traditional fishermen. *Fishery Technology,* **30** (1) : 154-158.

Chandargi, D.M. 1986. Extent of research and influence of motion picture entitled mixed farming on farmers – A field experiment, M.Sc. (Ag.) thesis (Unpub.), U.A.S., Dharwad.

Chandrakala, H.T. and Eswarappa, G. 2001. Knowledge and adoption of dairying practices of farm women in relation to their socio-personal characters, *Karnataka J. Agril. Sci.* **14** (1) : 95-100.

Chattopadhyay, S.N. 1963. A study of some psychological correlates of adoption of innovations in farming, unpublished Ph.D. thesis, Division of Agricultural Extension, Indian Agricultural Research Institute, New Delhi.

Chaudhuri, H., Chakraborti, R.D., Sen, P.R., Rao, N.G.S. and Jena, S. 1975. A new high in fish production in India with record yields by composite fish culture in fresh water ponds. *Aquaculture,* **6** : 343-55.

Chaudhuri, H., Rao, N.G.S., Saha, G.N., Rout, M. and Kanaujia, D.R. 1978. Record fish production through intensive fish culture. *J. Inland Fish Soc. India.* **10** : 19-27.

Chede, P.M. 1988. A study of the constraints in adoption of selected dairy technologies by dairy farmers. M.Sc. (Ag.) (Unpub.) thesis, Punjab rao Krishi Vidyapeeth, Akola.

Chinnappareddy, B.V. 1993. Investment in soil and water conservation. an analysis of its impact in the Kalyanakere Watershed Project. Ph.D. thesis (Unpub.), U.A.S., Bangalore.

Chitambar, J.B. 1973. Introducing Rural Sociology pp 93-97, 5th Ed. Chap. 5, Wiley Eastern Ltd., New Delhi.

Chitambar, J.B. 1981. Tools of understanding from the field of cultural anthropology: Introductory Rural Sociology, pp. 110.

Colekman, J.C. 1977. Introductory psychology, pp. 298-320, 1st Ed., Chapter 11, Section III. Routledge and Kegan Paul Ltd., London.

Copp, J.H. 1958. Toward Generalization in Farm Practices Research. Rural Sociology, **23** : 103-111.

Dana, S.S. 1987. A study of factors affecting the attitude of livestock-owners towards artificial insemination in cattle. M.V.Sc. thesis (Unpub.), IVRI,Izatnagar,U.P.

Das, P, Bhaumik, U, Pandit, P.K., Roy, B., Banerjee, B.K. and Mondal, S.K. 1988. Some variables contributing to the adoption of composite fish culture innovations. In : M. Mohan Joseph (Ed.). The First Indian Fisheries Forum Proceedings, Asian Fisheries Society, Mangalore, pp. 467-470.

Day, F. 1873. Report on the fresh-water fish and fisheries of India and Burma. Calcutta, Superintendent of Government Printing, 307p.

De, K.C. 1910. Report on the fisheries of eastern Bengal and Assam. Shillong. 77p.

Dean, A., Aurbach, H.A. and Marsh, P.E. 1958. Some factors related to rationality in decision making among farm operators. Rural Sociology. **23** : 125-135.

Dehadrai, P.V. 2003. Indian aquaculture in 21st Century. Fish for All : National Lunch held at Kolkata, India, December 18-19, 2003 organized by ICLARM, Govt. of India and W.B. and M.S. Swaminathan Foundation, pp. 1-11.

Deway, R. and Humber, W.J. 1966. An Introduction to Social Psychology. New York, McMillan Company.

Digrashkar, R.K. and Wangikar, S.D. 1990. An investigation into the attitude of the farmers towards the use of biogas plant. *Maharashtra J. Extn. Edu.* **12** : 335-339.

Dixit, S. 1988. A study on the impact of social forestry programme on rural beneficiaries, M.Sc. (Ag.) thesis (Unpub.), U.A.S., Bangalore.

Dunsford, H.S. 1911. A report on the fisheries resources of the Punjab.

Ebel, R.L. and Damrin, D.E. 1960. Tests and Examinations Encyclopedia of Educational Research Ed. Harris C.W., The MacMillan Company, New York.

Edwards, A.L. 1969. Techniques of attitude scale construction. Vakils and Simon Pvt. Ltd., Bombay.

English, H.B. and English, A.C. 1958. A comprehensive dictionary of psychological and psychoanalytical terms. Longman's Green and Company, New York.

Ernest, R.S. 1973. A study of communication utilization behaviour of small and big farmers and its implication to communication strategy. Unpublished Ph.D. thesis, Indian Agricultural Research Institute, New Delhi.

Eyde, E.H.H. 1923. Report on the fisheries of the United Provinces. Allahabad.

Fisher, R.A. and Yates, F. (1963). Statistical tables for biological, agriculture and medical research. Longman Group Ltd.

Fowler, Jr. F.J. 1989. Survey Research Methods. California. Sage publication.

Gautam, U.S., Chand, R. and Singh, D.K.(2007). Socio-personal correlations for decision making and adoption of dairy practices. *Ind. Res. Jr. Ext. Edu.* 7(2 and 3). May and Sept., 2007.

Ghosh, S.D., Pal, S, Mondal, A and Ghosh, A, 2002. Upliftment of socio-economic conditions of East Kolkata Wetland based population by non-conventional recycling of waste water through pisciculture. Paper presented at National Symposium on Fisheries enhancement in inland waters – challenges ahead, 27-28 April 2002, CICFRI, Barrackpore, West Bengal.

Goswami, M. and Sathiadhas, R. 2000. Fish farming through community participation in Assam, *Naga, The ICLARM Quarterly.* **23** (3) : July-Sept., 2000 : 29-32.

Govinda Gowda, S., Lakshminarayan, M.T., Krishnamurthy, B. and Ranganath, A.D. 2000. Attitude of farmers towards dairy management practices. *Curr. Res.* **28** : 45-46.

Govinda Gowda, S., Venkata Reddy, T.N., Shivananda Murthy, H. and Vastar, C.S. 1991. Socio-economic attributes of inland fishermen. *Indian J. Extn. Edu.* **27** (1 and 2) : 110-113.

Gowda, N. 1992. Consequence of watershed development programme – An analysis of Chitravali watershed project in Karnataka, Ph.D. thesis (unpub.), U.A.S., Bangalore.

Graven, R. 1963. The family role in decision making. *Journal of Cooperative extension*, 1(1):23-31

Gupta *et al.*, 1999. Integration of aquaculture into the farming systems of the flood prone ecosystems of Bangladesh. An evaluation of adoption and Impact. ICLARM Contribution No. 1506, Philippines.

Gupta, C.L. 1976. A study of differential motives of dairy farmers of Milk Co-operative Societies of Intensive Cattle Development Project, Karnal (Haryana) towards Dairy Innovations. Unpublished Ph.D. thesis, Division of Diary Extension, N.D.R.I., Karnal (Haryana), Punjab University, Chandigarh.

Gupta, K.G. 1908. Report on the results of enquiry into the fisheries of Bengal and into fishery matters in Europe and America. Calcutta. 112p.

Haque, M.A. 1989. Composite fish farming technology for rural development. In : Ray, G.L. (Ed.), Agricultural Technologies for Rural development, Vikas Publishing House Pvt. Ltd., New Delhi.

Haque, M.A. and Ray, G.L. 1983(a). Factors related to the adoption of recommended species of fish in composite fish culture. *Indian J. Extn. Edu.* **18** (1 and 2) : 74-83.

Haque, M.A. and Ray, G.L. 1983(b). Fish farmers perception of the problems in composite fish culture and measures suggested by them for increasing the yield of fish. *Indian J. Extn. Edu.*, **23** (1 and 2) : 42-46.

Hay, D.G.1951. Social participation of individuals in four rural communities of North -East. Rural sociology, 16: 127-37.

Heady, B., Hildreath, R.J. and Dean, G.W. (1937). Uncertainty, expectations and investment decision for a sample of Central Iowa Farmers. Ames Iowa Agricultural Experiment Research bulletin: 447.

Henry, E., Gamett, R.S. and Wood Worth, D. 1973. Statistics in Psychology and Education, Pub. by Arun Vakil and Sons, Bombay, India, pp. 351-356.

Hora, S.L. 1943a. Eradication of predatory fishes from tanks. *Indian Fmg.*, **4** (7) : 337-8.

Hora, S.L. 1943b. Fish farmers calendar. Calcutta, Amrita Bazar Patrika, Puja number, 8p.

Hora, S.L. 1943c. Hints on fish culture in inland waters. *Indian Fmg.*, **4** (3) : 139-40.

Hora, S.L. 1943d. How to grow more fish. *Indian Fmg.*, **4** (4) : 179-81.

Hora, S.L. 1943e. Manuring of ponds and artificial feeding of fishes. *Indian Fmg.*, **4** (11) : 559-62.

Hora, S.L. 1943f. Observations on the fisheries of the improved tanks in Birbhum district. Calcutta, Department of Fisheries.

Hora, S.L. 1943i. Vegetation in fishery tanks. *Indian Fmg.*, **4** : 179-81.

Hora, S.L. 1943j. Why and how to grow more fish. Calcutta, Department of Fisheries, 3p.

Hora, S.L. 1945d. Selection, preparation and management of fish nurseries. *Indian Fmg.*, **6** (9) : 408-9.

Hora, S.L. 1951. Fish culture in rice fields. *Curr. Sci.,* **20** (7) : 171-3.

Hornby, A.S. 2005. Oxford Advanced Learner's Dictionary. pp. 326.

Hornell, J. 1920. Administrative report of the Department of Fisheries, Madras, for the year 1918-19. *Madras Fish. Bull.,* **12** : 1-42.

Hutton, J.H. 1963. Census of India, 1931. J. India, Part-I- Report.

Ingle, P.O., Rajkumar, R.N. and Rahad, B.G. 1991. Adoption behaviour of fishermen, *Maharashtra J. Extn. Edu.,* **10** (2) : 178-184.

Jaiswal, N.K. 1965. A study of factors associated with low level of adoption of improved agricultural practices. Unpublished Ph.D. thesis, Indian Agricultural Research Institute, New Delhi.

Jhingran, V.G. 1997. Fish and Fisheries of India, Hindustan Publishing Corporation (India), New Delhi, pp. 276.

Jimmiel, J.M. 1995. Socio-economic factors that influence the adoption of small scale rural fish farming at household level in Zimbabwe, *NAGA, The ICLARM Quarterly,* **18** (No. 2) : 25-28.

Juliana, W. 1964. Decision making in middle Sweden's farming, A special behavioural analysis. World Agricultural Economics and Rural Sociology Abstract, **6** : 808.

Kagan, J. and Havemann, E. 1976. Psychology – an introduction, pp. 490-509, 3rd Ed., Chapter 4. Harcourt Brace Jovanovich Inc., New York.

Kapse, S.S. 1976. Adoption of selected dairy husbandry practices by cattle owners in the Intensive Cattle Development Area of Dhulia (Maharastra State). Unpublished M.Sc. (Dairying) thesis, Division of Dairy Extension, National Dairy Research Institute, B.N.C. University, Kurukshetra (Haryana).

Kelly, E.L. 1967. Assessment of human characteristics, pp 19-20, 1st Ed. Chap-2, Prentice Hall of India Private Ltd., New Delhi.

Kluckhohn, H.P. 1965. Values and value orientation in theory of action. In 'Towards a general theory of action'. Edited by Parson T. and H.A. Shils. Massachusetts, Harvard University Press, Cambridge.

Krishnaiah, N.V. 1989. A study on effectiveness of short duration training programme conducted by FFDA in Andhra Pradesh. M.Sc. (Ag.) thesis (Unpub.), ANGRAU, Hyderabad.

Krishnamurthy, K. 1997. A study on knowledge and adoption of improved dairy management practices by farmers trained by Rudsell, Chitradurga. M.Sc.(Ag.) thesis (unpub.), U.A.S., Bangalore.

Krishnan, N. 1988. Attitude of fishermen towards fishermen Cooperative Societies. M.Sc. (Ag.) thesis (unpub.), Agril. College and Research Institute, TNAU, Madurai.

Kumar, D., Bhaumik, U. and Das, P. 1979. Production-cum-demonstration of composite fish culture in West Bengal. In : Symposium on Inland Aquaculture (Abstract), February 12-14, 1979, 69-70, CIFRI, Barrackpore.

Kumar, M.V. 1976. A study of adoption of few selected dairy practices in upland and delta areas of Vijayawada Taluk (Krishna District, A.P.). Unpublished M.Sc. (Dairying) thesis, Division of Dairy Extension, National Dairy Research Institute, Karnal, submitted to B.N.C. University, Kurukshetra (Haryana).

Kumar, D. (1992). Fish culture in undrainable ponds -A manual for extension. FAO technical paper 325, FAO, Rome.

Kunzru, O.N. and Tripathi, H. 1994. A comparative study of adoption of dairy farm technologies between non-member and member of dairy cooperative village. *Indian J. Animal Sciences,* **64** (5) : 501-507.

Le Xuan Sinh, 1996. The effects of aquaculture on farm household economy : A case study in Omon district, Cantho Province, Vietnam, *NAGA, ICLARM Quarterly,* January 1995.

Leagans, J.P. 1961. The communication process. Extension Education in Community Development. Published by Directorate of Extension, Ministry of Food and Agriculture, Government of India, New Delhi.

Likert, R. 1932. A technique for the measurement of attitudes. *Archeology Psychology,* No. 140.

Lindgren, H.C. and Byrne, D. 1975. Psychology – an introduction to behavioural science, pp. 502-05, 4th Ed., Chapter 15, John Wiley and Sons, INC; New York.

Lionberger, H.F. 1960. Adoption of new ideas and practices. Ames, Iowa State University Press.

Mahandra Kumar, K. 1996. Communication behaviour of fish farmers in Tamil Nadu, Ph.D. thesis, CIFE, Mumbai.

Makkar, S.L. and Shoal, T.S. 1974. Association of certain personal characteristics of farmers with their attitude towards soil testing. *Indian J. Extn. Edu.,* **10** (3 and 4) : 50-51.

Mallick B. 2001. Status of fisheries and socio-economic conditions of fisher folk around the Chilka lake of Orrisa. M.F.Sc. thesis (unpub.), CIFE, Mumbai.

Mande J.V., Rajput R.D., Thombare B.M. (2008). Knowledge of cattle owners about improved cattle rearing practices. J. dairying, Foods and Home Sc. 27(1):38-42, 2008.

Manjula, N. and Belli, R.B. 1994. Attitude of farm women towards watershed management programme. *Maharashtra J. Extn. Edu.,* **13** : 227-230.

Mankar, D.M., Sawant, V.Y. and Sashastrabudhe, A.G. 2000. Socio-economic and psychological profile of fishermen. *Maharashtra J. Extn. Edu.,* **18** : 159-163.

Mathew, P.M., Singh, B.K. and Chakraborty, D.P. 1979. Stocking of fry in composite fish culture. In : Symposium on inland aquaculture, February 12-14, 1979, CIFRI, Barrackpore, pp. 44.

Mathiyalagan, P. 1997. Acceptance pattern of poultry farm practices. *Maharashtra J. Extn. Edu.,* **16** : 100-103.

Mathiyalagan, P. and Subramanian, R. 1995. Extent of adoption of improved poultry practices. *Journal of Extension Education,* **6** (2) : 1130-1133.

Meeran, N.M. 1983. A study on the adoption behaviour of fish farmers, M.Sc. (Ag.) thesis (unpub.), Agricultural College and Research Institute, Madurai, TNAU.

Meeran, N.M. 2000. Influence of socio-personal, socio-economic and socio-psychological characteristics on the adoption behaviour of shrimp farmers. *Journal of Extension Education,* **11** (2) : 2742-46.

Meeran, N.M. 2001. Factors influencing the yield obtained by shrimp farmers. *Journal of Extension Education,* **11** (4) : 2937-2941.

Meeran, N.M. and Jayasheelan, M.J.P. 1996. Perceived problems of shrimp farmers. In : The Proceedings of the Fourth Indian Fisheries Forum, Nov. 96, pp. 495-97.

Meti, S.K. 1998. Technology research, perception, knowledge, adoption and attitude towards agricultural technologies by small and marginal farmers in TBP Area, Ph.D. thesis (unpub.), U.A.S., Dharwad.

Mishra, B.P. and Mishra, R. 1999. Constraints and scope of integrated fish farming. *Intensive Agriculture,* **37** (5-6) : 9-10.

Misra, S.R. 1987. Fisheries in India, Ashis Publishing House, New Delhi, pp. 84.

Mohammed Rafi, B. 1993. Knowledge, attitude and adoption of farm forestry among the participant and non-participant farmers in Shimoga district. M.Sc. (Ag.) thesis (unpub.), U.A.S., Bangalore.

Moses, S.T. 1947. Baroda Fisheries. *Bull. Dep. Fish. Baroda,* **11** : 1-10.

Moulik, T.K. 1965. A study of the predictive values of some factors of adoption of nitrogenous fertilizers and the influence of sources of information on adoption behaviour. Ph.D. Thesis (Unpub.), IARI, New Delhi.

Moulik, T.K. and Rao, C.S.S. 1973. Self-rating personality scale for farmers. In : Pareek, U. and Rao, V.T., Handbook of psychological and social instruments, Samasthi, Baroda.

Mulay, S. and Sabarathnam, V.E. 1980. Research methods in Extension Education. Archan Prakashan, New Delhi.

Mukundan, K. 1987. Modernization of prawn culture – suggested policy changes for the improved farming system in Pokkali areas – Kerala State, *Rural Development Review,* **6** (3 and 4) : 593-598.

Munda, B.S., Singh, R; Tewari, R and Sontakki, B. 1997. Gender based training needs for fisher folk in Mayurbhanj district, Orissa. *Journal of Extension Education,* **8** (2) : 1695-96.

Munn, N., L. Jr. Fernald, L.D. and Fernald, P.S. 1969. Introduction to psychology, pp 606-608, 3rd Ed. Chap 21, Oxford and IBH Publishing Co., New Delhi.

Nagaraj, K.H. 1999. An analysis of yield gap, technological and constraints in groundnut production, Ph.D. thesis (Unpub.), U.A.S., Bangalore.

Nagaraj, K.H., Lalitha, B.S. and Lalitha, K.C. 2000. Relationship between selected characteristics of groundnut growers and their adoption towards groundnut technology. *Curr. Res.,* **29** : 29-30.

Nagarajaiah, C.S. 2002. A study on knowledge, attitude and extent of adoption of composite fish culture practices in South Karnataka. Ph.D. thesis (Unpublished). Central Institute of Fisheries Education, Versova, Mumbai-400061.

Naidu, M.R. 1939. Report on a survey of the fisheries of Bengal. Calcutta, Superintendent Government Printing, 88p.

Nair, G.T. 1969. A multivariate study on adoption of high yielding paddy varieties by the farmers of Kerala State. Unpublished Ph.D. thesis, Indian Agricultural Research Institute, New Delhi.

Napier, T, Cameron, S., Thraen and Silvanna, M. Combini, S 1988. Willingness of land operator to participate in Government sponsored soil erosion control programmes. *J. Rural Studies.* **4** (4) : 339-347.

Natarajan, K. and Muthaiah, M. 1995. Attitude of farmers towards social forestry programme. *Journal of Extension Education,* **6** (1) : 1089-1090.

Nataraju, M.S. and Chande Gowda, M.B. 1984. Differential adoption of improved dairy practices by small farmers, marginal farmers and agricultural labourers. *Indian J. Extn. Edu.,* **20** (3 and 4) : 63-66.

Nayak, A.K., Behera, B.K., Misra, C.K., Mukhi, S.K. and Kohil, M.P.S. 2001. Constraints in the adoption of shrimp farming in Orissa. *Applied Fisheries and Aquaculture 2001.* **1** : 111-112.

Nelson, L., Ramsey, C.F. and Vernon, C. 1965. Community structure and change. The MacMillian Co., New York.

Newcomb, T.M., Turner, R.H. and Converse, D.E. 1965. Social Psychology : A study of human Interaction. New York. Holt Rinehart and Winston, Inc.

Nicholson, F. 1915. Papers from 1899 relating chiefly to the development of Madras Fisheries Bureau. *Madras Fish. Bull.,* **1** : 237p.

Nicholson, F. 1918. Annual reports of the Madras Fisheries Bureau 1908-1917. *Madras Fish. Bull.,* **10** : 179p.

Nijagonda, H. 2000. A study on attitude, communication and marketing pattern of red gram growers in Bidar district, M.Sc. (Ag.) thesis (Unpub.), U.A.S., Bangalore.

Nimbal, M.F.K. 1995. A study on attitude knowledge of biogas technology user, M.Sc. (Ag.) thesis (unpub.), U.A.S., Dharwad.

Norris, T.L. 1953. Decision making activity sequence in a Harienda Community. Human Organization, **12** (3) : 26-30.

Panigrahi, A.K. and Dana, S.S. 2002. Extension strategies for sustainable yield in aquaculture sector, paper presented at National Symposium on Fisheries Enhancements in inland waters – Challenges ahead, 27-28 April, 2002, CICFRI, Barrackpore, West Bengal, India.

Panjabi, N.K. Mohammad, A. and Punia, M.K. 1999. Fish production in Udaipur district. *Agricultural Extension Review,* **11** (5) : 29-31.

Parsons, T. and Shils, E.A. 1965. Towards a General Theory of Action. Harvard University Press, Cambridge, Massachusetts.

Patil, S.L. 1990. A study on the extent of adoption of recommended cultivation practices of CDH32 Cotton by farmers and its economics in Malaprabha Command Area of Dharwad district. M.Sc. (Ag.) thesis (Unpub.), U.A.S., Dharwad.

Perumal, G., Ponnappan, C. and Balasubramaniam, S. 1988. Attitude of fish farmers towards Fish Farmers Development Agency (FFDA), *Rural Development Reviews, TNAU,* **7** : 667-668.

Petharam, R.J. 1985. Constraints to the transfer of Australian farming system Research for Developing centres. *Australian centre for International Agril Research,* Canberra, pp. 156-162.

Pochaiah, M and Reddy, P.R.K. 1998. A study on the extent of adoption and yield gap in paddy. *Journal of Extension Education,* **91** : 1925-1927.

Ponnappan, C. 1982. Fish Farmers Development Agency Programme : An analysis. M.Sc. (Ag.) thesis (Unpub.), TNAU, Coimbatore.

Ponnusamy, K., Gopinathan, K., Kumaran, M. and Krishnan, M. 2000. Constraints analysis in adoption of shrimp farming technology, paper presented at National conference on Fisheries economics, extension and management, CIFE, Mumbai, 5-6th June, 2000.

Pounraj, A. 1992. Inland Fishermen Cooperative Societies – A critical analysis, M.Sc. (Ag.) thesis, TNAU, Coimbatore.

Pounraj, A. and Sripal, K.B. 1997. Socio-economic characteristics of the member of inland fisheries Cooperative Societies. *Journal of Extension Education.* **8** (3) : 1789-90.

Pramod, K., Rout and Sudhir K. Das. 1992. Socio-economic aspects of extensive shrimp aquaculture along Gopalapur coast in Ganjam district of Orissa. *Sea Food Export Journal.* May 1992 : 5-8.

Prasad, M.S. 1995. Adoption of dry land agriculture technologies by farmers of Ranga Reddy district (AP). A Retrospective Analysis, Ph.D. thesis, U.A.S., Dharwad.

Prasad, M.S. and Sundara Swamy, B. 1999. Knowledge on dryland agricultural technologies. *Indian J. Extn. Edu.* **10** (4) : 2586-2591.

Praveena, I. 1993. A study on the constraints in adoption of fish culture as perceived by fish farmers of Nellore district of Andhra Pradesh. M.Sc. thesis, EEI, APAU, Hyderabad.

Radha Krishnan, T.V., Ravichandran, G. and Nirmala, L. 1977. Agricultural Extension Compendium, TANU, Coimbator, pp. 6.

Rahimullah, M. 1944. A plan for the development of fisheries in H.E.H. Nizam's Dominions. Hyderabad, Govt. Press, 24p.

Raj, B. 1916. Notes on the fresh-water fishes of Madras. *Rec. Indian Mus.*, **12** (6) : 249-94.

Raju Naik, H., Sundara Swamy, B. and Ansaraj, M.R. 1994. Attitude of farmers towards seed supplying agencies as influenced by their personal and socio-economic characteristics. *Maharashtra J. Extn. Edu.* **13** : 183-186.

Rajurkar, R.N. 1988. A study of adoption of improved fish farming practices by farmers. M.Sc. (Ag.) thesis (Unpub.), Punjab rao Krishi Vidyapeeth, Akola.

Ramachandran, P.K. 1974. A multivariate study on information source utilization of big, medium and small farmers. Unpublished Ph.D. thesis, Indian Agricultural Research Institute, New Delhi.

Ramesh Kumar Reddy, P. and Sivashankara Rao, V. 1998. Assessing knowledge, attitude and adoption of farmers about improved production in cabbage. *Maharashtra J. Extn. Edu.* **17** : 79-83.

Ramsay, C.E., Polson, R.A. and Spencer, G.E. 1954. Values and adoption of practices. Rural Sociology. pp. 24-47.

Ranadhir, M., Mitra, P.M., Sen, H.K., Das, P. and Bhowmik, U. (Comp). 1979. Additional 126 case studies of composite fish culture in India incorporating second batch of studies performed during 1974-78 in different part of the country. *Bull. Cent. Inland Fish. Res. Inst.* No. **29**, pp. 126 (mimeo).

Ranganathan, V. 1961. A review of chanos fry transport. In : Training Course on Fish Fry Transport, June 5th to 10th, 1961. Mandapam Camp, Fisheries Extension Unit, Government of India, 2p.

Rao, R.N. 1968. A study on preferential adoption of package of practices for high yielding varieties of rice in Andhra Pradesh. Unpublished M.Sc. Thesis, Division of Agricultural Extension, Indian Agricultural Research Institute, New Delhi.

Reddy, K.V. 1987. Adoption behaviour and information sources of farmers in respect of selected dairy practices. *J. Res. APAU*, **15** (1) : 39-45.

Reddy, M.S.K. and Kivlin, J.E. 1968. Adoption of high yielding varieties in three Indian villages. National Institute of Community Development, Hyderabad.

Reddy, P.V. 1997. A study on the entrepreneurial characteristics and farming performance of fish farmers in Nellore district of Andhra Pradesh, Ph.D. thesis (Unpub.), ANGRAU, Hyderabad.

Reiss, F. 1952. Individual differences in entrepreneurial and management ability among Illinois farm operators. Unpublished Ph.D. thesis, Illinois, Department of Agricultural Economics, University of Illinois.

Rogers, E.M. 1962. Diffusion of Innovations. The Free Press, New York.

Rogers, E.M. and Shoemaker, F.F. 1971. Communication of innovations, Edn. 1. The Free Press, A division of MacMillan Publishing Co. Inc., New York.

Rogers, E.M. and Svenning, L. 1969. Modernization among peasants. The impact of communication, Holt Rinehart and Winston. Inc. New York.

Roy, L. and Pathak, M. 2000. Attitude of farm women towards post harvest activities. *Indian J. Extn. Edu.* **36** (1 and 2) : 82-84.

Roy, P., Fliegel, F.C., Kivlin, J.E. and Sen, L.K. 1968. Agricultural innovations among Indian farmers. National Institute of Community Development, Hyderabad.

Rusdi, T. 1995. Source of information knowledge and adoption of technology by ornamental fish farmers in Jakarta, Indonesia. M.Sc. (Ag. Extn.) thesis (Unpub.), U.A.S., Bangalore.

Sah, A K and Chand, R. 1999. Adoption of improved dairy practices in Bihar, *Indian J. Extn. Edu.* **35** (3 and 4) : 234-236.

Santosh, K.D. and Sudhir, K.D. 1992. Economic analysis of fish production under extensive aquaculture practices in Ganjam district of Orissa. *Sea Food Export Journal,* October 1992 : 21-23.

Sarangi, N. 2006. Controlled seed production. Survey of Indian Agriculture 2006. The Hindu.

Sarkar, U., Islam, S., Goswami, A. and Mazumdar, D. (2007). Some Socio economic, Socio Psychological and Communication characteristics on adoption behaviour of the Rajbanshi Dairy farmers. journal of Interacademicia. 11(2): 230-237.

Savithri, A. 1992. A study of knowledge, attitude and adoption of dryland technologies by farm women – An expert factor analysis. M.Sc. (Ag.) thesis (Unpub.), TNAU, Coimbatore.

Seenappa, D. and Surendra, H.S. 1988. Knowledge and attitude of fishermen trainees towards fish culture. In : M. Mohan Joseph (Ed.) : The First Indian Fisheries Forum Proceeding, Asian Fisheries Society (Indian Branch), Mangalore, 463-468.

Sen, D. and Das, P.K. 1986. Problems of transfer for inland fisheries technology among the small scale fish farmers. *J. Rural Dev.* **5** : 313-359.

Sen, D.N. 1941. Observations on the method of carp culture in the so-called salt lakes near Calcutta with a note on the fauna of the lakes. *J. Roy. Asiat. Soc., Bengal,* **7** : 7-13.

Sengupta, J. 1967. A simple adoption scale for selection of farmers for high yielding varieties programme on rice. *Indian J. Extn. Edu.* **3** : 107-115.

Shankariah, C. 1965. A study of differential characteristics of vegetables and non-vegetables growers and factors associated with adoption of improved practices of vegetables cultivation. Unpublished M.Sc. Thesis, Division of Agricultural Extension, Indian Agricultural Research Institute, New Delhi.

Sharma, A. 2002. Study on the problem perceived by women fisheries towards participation in aquaculture activities for upliftment of socio-economic status. Paper presented at National Symposium on Fisheries enhancements in inland waters-challenges ahead. 27-28 April, 2002, CICFRI, Barrackpore, West Bengal.

Sharma, B.K. 1985. Strategy for transfer of composite fish culture practices. Lectures notes on composite fish culture and its extension in India. Fresh water training centre, Asia Bankok, Thailand, FAO.

Sharma, O.P. and Kumar, Anup 2001. Socio-economic status of fishermen in relation to fisheries of lake Jaisamand. *Fishing Chimes.* **20** (12) : 45-47.

Shashikumar, S. 1998. Performance of non-governmental organizations in rural development and people's attitude towards NGO's in Karnataka : A diagnostic study. Ph.D. thesis (Unpub.), U.A.S., Dharwad.

Sheela, B. 1991. A study on knowledge and adoption of improved dairy practices by women in Bidar district, Karnataka. M.Sc. (Ag.) thesis (unpub.), U.A.S., Dharwad.

Sheoran, V.K. and Ram Kumar, S. 1988. Correlates of adoption of dairy innovations of IRDP beneficiaries. *Indian J. Extn. Edu.* **24** (1 and 2) ; 71-73.

Sherif, M. and Cantrill, H. 1948. The psychology of attitudes. Psychological Review, **52** : 295-319.

Shinde, P.S., Bhople, P.P. and Chikhale, N.J. 1999. Multivariate analysis of adoption of animal husbandry management practices by dairy farmers. *Maharashtra J. Extn. Edu.* **18:** 111-113.

Shinde, V.G., Sangle, G.K. and Dikle, R.N. 1998. Factors associated with adoption of dairy practices by farmers. *Maharashtra J. Extn. Edu.,* **17** : 108-116.

Shivashankarappa, 1995. Impact of watershed development programme on farmers of North Karnataka. Ph.D. thesis (Unpub.), U.A.S., Dharwad.

Siddaramaiah, B.S. and Reddy, H.C.V. 1993. Attitude and adoption behaviour of farmers towards watershed management. *Mysore J. Agric. Sci.* **27** (3) : 291-296.

Sidhu, K.S. 1980. A study of dimensional variation in the knowledge of dairy farmers of I.C.D.P. and non-I.C.D.P. area of Punjab. M.Sc. thesis (unpublished), Kurukshetra University, Kurukshetra.

Sigman, V.A. and Swamson, B.E. 1984. Agricultural extension-A reference manual, second edition. Swanson (Ed.) FAO, Rome, pp-227-231.

Simon, H.A. 1957. Models of man, social and rational. Mathematical essays on rational and human behaviour in a social setting. New York, John Wiley and Sons.

Singh, B.N. and Jha, P.N. 1965. Utilization of sources of farm information in relation to adoption of improved agricultural practices. *Indian J. Extn. Edu.* **1** (1) : 34-42.

Singh, R. and Sohal, T.S. 1967. Size of holdings as related to acceptance of crop plans. Extension contacts and education of the farmers. *Indian J. Extn. Edu,* **3** (1-2) : 42-48.

Singh, S.N. 1969. A study on adoption of high yielding varieties and investment pattern of additional income by the farmers of Delhi Territory. Unpublished Ph.D. thesis, Indian Agricultural Research Institute, New Delhi.

Singh, S.N. and Singh, K.N. 1970. A multivariate analysis of adoption behaviour of farmers. *Indian J. Extn. Edu.* **6** (3-4) : 39-45.

Singh, T.R. 1975. Analysis of rationality in decision making in relation to the adoption of innovations of varying nature. Unpublished Ph.D. thesis, Indian Agricultural Research Institute, New Delhi.

Singh, B and Kunzru, O.N. 1985. Attitude of farmers towards goat and sheep rearing. *Indian J. Extn. Edu,* **21** (1 and 2) : 86-88.

Singh, R. and Nande, 2001. Training needs of fish farmers. *Journal of Extension Education,* **12** (1) : 3064-65.

Sinha, V.R.P. and Gupta, V.M. 1975. On the growth of grass carp *Ctenopharyngodon idella* var. in composite fish culture at Kalyani, West Bengal (India), *Aquaculture.* **5** : 283-90.

Sinha, I.M. 1978. A study of some selected factors associated with adoption of chemical fertilizers. Unpublished M.Sc. (Ag.) thesis, Division of Agricultural Extension, Rajendra Agricultural University, Ranchi Campus, Bihar.

Sinha, P.R.R. and Kolte, N.V. 1974. Adult education in relation to agricultural development – An evaluation study of a development block in Andhra Pradesh, NIRD, Hyderabad.

Sinha, V.R.P. 1987. In : Fish and Fisheries of India, Hindustan Publishing Corporation (India), Delhi, pp. 442.

Sinha, V.R.P., Gupta, V.M., Banerjee, M.K. and Dhirendra Kumar, M 1973. Composite fish culture at Kalyani, West Bengal. *J. Inland Fish Soc. India.* **5** : 201-07.

Somasundram, D. and Singh, S.N. 1978. Factors affecting the knowledge of adopter and non-adopter small farmers. *Indian J. Extn. Edu.* **14** (1 and 2) : 30-34.

Srinath Krishna, 1988. Awareness and attitude of prawn farmers towards scientific prawn farming. *Indian J. Extn. Edu.,* **26** (1 and 2) : 74-76.

Srinath, K. 1996. Adoption of shrimp culture practices by farmers. *Sea Food Export Journal,* October 1996 : 9-12.

Srivastava, U.K., Dholakia, B.H., Sreenivas Rao, S. and Vathsala, S. 1993. Freshwater aquaculture in India, Oxford and IBH Publishing Co. Pvt. Ltd. New Delhi, pp. 374.

Subashini, B. and Thyagarajan, S. 2000(a). Socio-personal characteristics of Tapioca growers. *Journal of Extension Education,* **11** (1) : 25-26.

Subashini, B. and Thyagarajan, S. 2000(b). Socio-personal characteristics of Tapioca technologies. *Journal of Extension Education,* **36** (3 and 4) : 153-155.

Subhash Chandra, R. 1986. Consequences of adoption of fish culture practices by fish farmers, M.Sc. (Ag.) thesis (Unpub.), TNAU, Coimbatore.

Sujath Kumar, N.V. 1988. Adoption behaviour of traditional fishermen and trawler owners – A comparative analysis. M.Sc. (Ag.) thesis, TNAU, Coimbatore.

Sukumaran, K.K. 1976. Progress report of the Karnal Sub-Centre of Haryana for the period Nov. 72 to Aug. 75, In : Third Workshop on the All India Coordinated Research Project on Composite Fish Culture and Fish Feed Production, February 26-27, 1976, CIFRI, Puri.

Sumathi, P. and Alagesan, V. 2000. Level of adoption of IPM practices by groundnut growers. *Journal of Extension Education,* **2** (1) : 9-12.

Sunder Rao, D. 1988. Impact of improved dryland agricultural technology in Chevella and Pothulaboguda model watershed development project in Jogipal Taluk of Medak district in Andhra Pradesh. M.Sc. (Ag.) thesis (Unpub.), APAU, Hyderabad.

Supe, S.V. 1969. Factors related to different degrees of rationality in decision making among farmers. Ph.D. thesis (Unpub.), IARI, New Delhi.

Supe, S.V. 2002. Introduction to Extension Education. 4th edn. Chapter 2nd, Oxford and IBH Publishing Co. Pvt. Ltd., New Delhi.

Supe, S.V. and Singh, S.N. 1973. Dynamics of rational behaviour of Indian farmers. New Heights and Publishers and distributors, Darya Ganj, New Delhi, pp. 211-117.

Surekha, S., Pemalatha, B. and Nagaraja, N. 1997. Attitude of farmers towards watershed development. *Maharashtra J. Extn. Edu.,* **16** : 82-85.

Suresh, R., Selvaraj, P. and Vasanthakumar, J. 1988. Yield gap and constraints in inland fish culture – A micro level study. In : M. Mohan Joseph (Ed.), The First Indian Fisheries Forum, Proceedings, Asian Fisheries Society, Indian Branch, Mangalore, pp. 439-440.

Suresh, R., Sukumaran, M. and Selvaraj, P. 1990. An economic analysis of productivity in freshwater aquaculture in Madhurai district. The Second Indian Fisheries Forum Proceedings, May 1990, Mangalore, India, pp. 345-347.

Sushant P, Oomanchan, L. and Khare, N.K. 2002. Knowledge and attitude of members of fishermen cooperative society towards Trapa-cum-fish farming in Jabalpur district (MP), enhancement in inland waters, challenges ahead. 27-28th April 2002. CICFRI, Barrackpore, West Bengal.

Swathi Lakshmi, P.S. 1995. Social forestry – A diagnostic analysis. M.Sc. (Ag.) thesis (Unpub.), TNAU, Coimbatore.

Talukdar, P.K. 2000. Knowledge level and extent of adoption of composite fish culture practices by aquaculturists in Sonitpur district of Assam. M.F.Sc. thesis, CIFE, Mumbai.

Talukdar, P.K. and Sonataki, B.S. 2005. Correlates of adoption of composite fish culture practices by fish farmers of Assam, India. *The Journal of Agriculture Sciences*, Vol. 1 (1).

Tawade, N.D., Patil, P.A. and Nirban, A.j. 1997. Aspiration of fishermen community about their living. *Maharashtra J. Extn. Edu.,* **16** : 143-146.

Thorat, S.S. 1969. Influence of traditional and non-traditional status on the adoption of health practices. Behavioural Sciences and Community Development, **3** (1) : 38.

Thorndike, R.L. and Hagen, E.P. 1979. Measurement and evaluation in psychology and education, pp. 395-396, 4th Ed., Chapter 11, Wiley Eastern Ltd., New Delhi.

Thornton, D.S. 1962. The study of decision making and its relevance to the study of farm management. The Farm Economist. **10** (2) : 40-55.

Thurstone, L.L. 1946. Comment, *Amer J. Social,* **52** : 39-50.

Toulin, S. 1974. Post Script, the structures of Scientific theories. University of Ilinois Press, Urbana, USA.

Tripathi, S.D. 1979. In : Fish and Fisheries of India. Hindustan Publishing Corporation (India), Delhi, pp. 436.

Tripathi, S.D. 2003. Inland aquaculture in India. Fish for All : National Lunch held at Kolkata, India, December 18-19, 2003 organized by ICLARM, Govt. of India and W.B. and M.S. Swaminathan Foundation, pp. 33-57.

Van der Mheen, 1999. Adoption of integrated aquaculture and irrigation. A study conducted in Zambia and Tanzania. TCP/RAF/7825, ALCOM Working paper no. 23, Rome, FAD, pp. 21.

Venkataramanujam, K.N., Sukumaran, R and Sanjeeviraj, G.1987. Socio-economic conditions of the Chank Fishermen Community in Tutuiconn (S. India). *Rural Development Review,* **6** (3 and 4) : 605-609.

Venkatesh Prasad, G. and Siddaramaiah, B.S. 2000. Correlates of plant protection measures. *Indian J. Extn. Edu.,* **26** (3 and 4) : 124-133.

Viju, A. and Pillai, G.B. 1988. A study on the attitude of Kanikkars towards farming and factors related with it. *Rural Dev. Review.* **7** (1 and 2) : 682-686.

Wadt. E. and Brown, G.W. 1963. Essentials of Educational Evaluations, Holt, Rinehart and Winston, New York.

Wagmare, S.K. and Pandit, V.K. 1982. Constraints in adoption of wheat technology by the tribal farmers. *Indian J. Extn. Edu,* **1**(1) : 70-71.

Wagmare, V. and Wagmare, S.K. 1987. Image and impact of transfer of technology. *Programme on small farmers,* Metropoliton Book G. pvt. Ltd. New Delhi.

Wayal, J.B. and Ingle, P.O. 1994. Fisheries development prospects in Akola. *Agricultural Extension Review,* March-April, pp. 18-22.

Wayal, Jagadeo Bhikaji, 1990. Study on prospects of fisheries development in Akola district. M.Sc. (Ag.) thesis, Punjab Rao Krishi Vidyapeeth, Akola.

Wilkening,.A. 1953. Adoption of improved farm practices as related to family factors. *Wisconsisn Agrico. Expt. Sta. Res. Bull.* pp. 183.

Wilkening, E.A. 1958. Communication and technological changes in rural society in Alyin A., Bertrand (Ed.) Rural Sociology. Analysis of Contemporary Rural Life. New York, McGraw Hill.

Wright, S. (1934). The Method of Path coefficients. Annals of Mathematical Statistics. 5: 161-215.

Yadav, Y. S. 2003. Responsible aquaculturists and aquaculture – securing a safe future for the aquaculture sector. Fish for All : National Lunch held at Kolkata,

India, December 18-19, 2003 organized by ICLARM, Govt. of India and W.B. and M.S. Swaminathan Foundation, pp. 73-82.

Zinyama, L.M. 1988. Farmers perception of the constraints against increased crop production of the constraints against increased crop production in the subsistence commercial farming sectors of Zimbabwe. Agric. Administration and Extension, pp. 97-107.

APPENDICES

Interview Schedule
on
Farmers and Scientific Fish Culture

Sl. No. :

Date :

PART-1

General Background information of fish farmers

Name:	*Village:*
Block :	*District :*
State :	*Total pond area :*
No. of pond:	*Experience in fish culture:*
Type of Fish culture (Semi intensive/modified extensive/traditional) :	
Annual production (qt.) of fish :	*Annual family consumption :*
Quantity sold/annum :	

PART-II

Independent Variables

A. Socio-personal Variables

Age :	*Sex :*	*Caste (Gen/SC/ST) :*	
Family size :		*Family type(Joint/Nucleus) :*	
Education :			
a) Illiterate	b) Can read and write	c) Primary	d) Middle
e) Secondary	f) H.S.	g) Graduate	h) Post-Graduate

B. Socio-Economic Variable

Occupation :	*Primary :*	*Secondary :*
Annual income :		*Income from fish culture :*
Land holding:		

Name of the holding	*Area*	*Irrigated*	*Rainfed*
Owned			
Leased in			
Leased out			
Total (a+b+c)			

Possession of Fishing Equipments

a) No equipments

b) Cast nets

c) Gill nets

d) Drag net

e) Coracle with gill net and drag net

f) More than one unit of coracle with nets

Social Participation

Are you a member of any of the following organization? (Yes/No):

If yes then tick mark (√) the organization and also indicate whether you are a member/office bearer and if so how frequently you attend the meetings?

Sl.No.	*Organisation*	*As Office Bearer*	*As Member only*	*Attend Meeting*		
				Regularly	*Occasionally*	*Never*
a.	Gram Panchayat					
b.	Panchyat Samiti					
c.	Zilla parisad					
d.	Rural youth club					
e.	Religious committee					
g.	Any other (specify)					

C. Communication Variables

a) Communicator facilities

Sl.No.	*Items*	*Available with the Respondent*	*Available with Neighbours*	*Not at Available*
1.	Radio			
2.	T.V.			
3.	Newspaper			
4.	Farm Magazine			
5.	Telephone			
6.	Any other			

b) Mass media participation

Sl.No.	*Mass Media*	*Owning*		*Subscription*		*Frequency*		
		Owning	*Not owning*	*Subscription*	*No Subscription*	*Regular*	*Occasional*	*Never*
1.	Listening to radio (General prog.)							
2.	Listening to Radio (Fisheries and Ag. prog.)							
3.	Listening to T.V. (General Prog.)							
4.	Listening to T.V. (Fisheries and Ag. prog.)							
5.	Reading newspaper							
6.	Reading farm magazine, leaflets etc.							

c) Cosmopoliteness

Please state your agreement/disagreement with each of the following statements.

Sl.No.	Statements	Response Categories		
		A	UD	DA
1.	An individual fish farmer can learn everything he needs only through experience in his own village			
2.	All the things that an individual needs to know for his own benefit are not necessarily found in his own village			
3.	All the needs of an individual can be entirely met within his own village			
4.	In the present age of increase transport and communication facilities an individual should know more about the thing happening outside the village			

d) Extension Agency Contact

Do you contact the following persons to collect information regarding scientific fish culture practices ? Yes/No

If yes, how often?

Sl.No	Persons	Frequently	Often	Seldom	Never
a.	Fishery Extension Officer				
b.	Fishery Assistant				
c.	Scientists (University/KVK)				
d.	Progressive fish farmer				
e.	Any other				

D. Psychological Variables

a) Innovative proneness

Sl.No.	*Statements*	*Response Categories*				
		SA	*A*	*UD*	*DA*	*SDA*
1.	I am very much interested in adopting whatever new practices that are helpful in better fish culture practices					
2.	Since, we are not sure of the new methods I would like to wait till others adopt					
3.	Since, new methods/improved techniques of fish culture are not profitable, I am not interested in any of them					
4.	I try to keep myself well informed about the improved practices and try to adopt as soon as possible					
5.	New practices/methods of fish culture are not easily adaptable and hence I do not adopt					

b) Credit orientation

Please indicate your response for the following:

i.	Do you think a farmer like you should borrow money for fish culture purposes.	Yes/No
ii.	In your opinion how difficult it is to secure credit in fish culture purposes.	Very easy/Easy/ Difficult/Very difficult
iii.	How a fish farmer is treated when he goes to secure credit?	Very fairly/Fairly/ Badly/Very badly
iv.	There is nothing wrong in taking credit from institutional sources for increasing fish production.	SA/A/DA/SDA
v.	Did you use credit in the last two years for fish culture?	Yes/No

c) Value Orientation

(Localite cosmopolite and external conformity individualism)

Value orientation of fish farmers differ. The following are some statements representing value orientation of fish farmers. Please state the degree of your agreement or disagreement with each of them, on the 4 point scale as follows :

Sl.No.	*Statements*	*Response Categories*			
		SA	*A*	*DA*	*SDA*
1.	A man can learn many things from the happenings and experience of his village only.				
2.	One's experience is not as better as collective experience of a group of fish farmers.				
3.	A person who has seen something worked in his village need not worry about taking any additional information from sources outside his village.				
4.	A person who does not believe in consulting other can do better job.				
5.	One can satisfy all his requirements out of the local resources available to him.				
6.	A person can save himself with many in decisions and difficulties of life, if he believes in taking and following the advice of his friends, relatives, neighbours etc.				
7.	Many things a person ought to know are happening outside of his village and such happening may be of great advantage to a person. He, therefore, ought to know.				
8.	It is a sign of weakness and impotency when a person relies on other's opinions for making his decisions.				
9.	At present when transport and other communication facilities are developing, a villager should know more about things happening outside his village.				
10.	Action should be undertaken only after consulting others.				

d) Risk orientation

Sl.No.	*Statements*	*Opinion*				
		SA	*A*	*U*	*D*	*SDA*
1.	Fish farmer should rear more number of species to avoid greater risks involved in growing one or two species.					
2.	A fish farmer should rather take more of a chance in making a big profit than to be content with a smaller but less risky profit.					
3.	A fish farmer who is willing to take greater risks than the average farmers, usually does better financially.					
4.	It is a good for a fish farmer to take risk when he knows his chance of success is fairly high					
5.	It is better for a fish farmer not to try new fishery practices unless most of other farmers have used them with success.					
6.	Trying an entirely new method in fish farming by a farmer involves risk but it is worth it.					

e) Economic motivation

Sl.No.	*Statements*	*Opinion*				
		SA	*A*	*UD*	*DA*	*SDA*
1.	A fish farmer should work towards more fish production and economic profit.					
2.	A most successful fish farmer is the one who makes the most profit.					
3.	A fish farmer should try any new technological know how of fish farmer which may earn him more money.					
4.	A fish farmer should adopt modern methods of production to produce more fish to increase monetary profits in comparison to traditional methods					

Sl.No.	Statements	Opinion				
		SA	A	UD	DA	SDA
5.	It is difficult for the children of fish farmer to make a good start unless he provides them with economic assistance.					
6.	A fish farmer must earn his living but the most important thing in life can not be defined in economic terms.					

E. Situational Variables

1. Name of the water body ________________; Location ________________; Distance from residence (km) ________________________
2. Ownership : Self owned/leased/auction
3. Main source of water : Rainfed/canal fed/both
4. Secondary source of water ____________________________________
5. Water retention : short seasonal/long seasonal/perennial(6-9 months) (9-18 months)
6. Average water depth (m) : ____________________________________
7. Period of lease (yrs.) __________________, from ________________ to ______________
8. Lease amount/year (Rs.) ____________________________________
9. Infested with weeds : Yes/No
10. Extent of weed infestation : completely choked/moderate/low/Nil
11. Type of weeds : Floating/Sub-marginal/Marginal/Emergent/All
12. Existence of local fauna : Nil/Scanty/Abundant
13. Varieties/types of local fauna : ____________________________________
14. Uses of water body : Irrigation ____________; Domestic ___________, Both ___________

15. In case of self owned pond : (1) App. Cost of excavation ____________
(2) Source of water ______________________(3)

Culture Practices

16. Type of culture practices : Monoculture/Polyculture – composite/ Integrated

17. Culture input level :
(a) Stocking only ( )
(b) Stocking and manuring ( )
(c) Stocking and feeding ( )
(d) Stocking, manuring and feeding ( )
(e) Stocking, manuring and liming ( )
(f) Stocking, feeding and liming ( )
(g) Stocking, manuring, feeding and liming ( )

18. Species stocked : Single/2 spp./3 spp./4 spp./5 spp./6 spp.

19. Name of species stocked : ______________________________________
____

20. Rate of stocking (No/ha) : __________________________ : Not fixed
____

21. Types of manure used : Cowdung/poultry/both/no manure/others (specify)

22. Source of manure : ____
____

23. Interval of manuring : Weekly/15 days/monthly/bimonthly/once in three months irregular (occasional)

24. Types of feed used : Rice bran and oil cake in recommended ratio/Rice bran and oil cake in other than recommended ratio/rice bran only/domestic waste/others (specify)

25. Source of feed : ____
____

26. Interval of feeding : daily/2-3 times per week/weekly/15 days/monthly/ irregular

27. Watch and ward : Regular/irregular/but erected at tank site/no hut/ self/hired/both

28. Method of harvesting : partial/complete/need based
29. Mode of harvesting : self/employing labour (fishing units)/both/royalty fishing/auction
30. Mode of disposal : self retailing/sold to merchant (wholesaler)/both/ through FCS/all the above

F. Knowledge Test on Scientific Fish Culture

Sl.No.	Practices	Score
1.	What in kind of soil is good for fish culture ?	Clay loamy/loamy-(1) Any other - (0)
2.	What in the minimum depth of water required for fish culture ?	1 m - (1) Any other - (0)
3.	What are the nutrients required for production of natural fish food organisms in fish pond ?	N/P - (1) Any other - (0)
4.	Is it necessary to use lime in fish culture ?	Yes - (1)No - (0)
5.	How do you correct acidic condition of fish culture pond/tank ?	Using lime - (1) Any other - (0)
6.	Do you know the recommended dosage of lime used in general ?	250-300 kg/ha/yr - (1) Any other - (0)
7.	Should we have to manure the fish culture ponds?	Yes - (1) No - (0)
8.	What are the advantages of manuring fish culture pond ?	Increases stocking rate, increase production of fish food organisms - (1) Any other - (0)
9.	Name some common organic manures used in fish culture	RCD/Poultry manure/ Pig dung-(1) Any other -(0)
10	Do you know the rate of application of cow dung (including initial dose and subsequent monthly doses) ?	RCD 5000 kg/ha/yr - (1) Any other - (0)
11.	How many days before of stocking of fish seed manure should be applied ?	7-10 or 15 days - (1) Any other - (0)
12.	Is it necessary to use inorganic fertilizers in addition to organic manures in fish culture ?	Yes - (1) No - (0)

Sl.No.	*Practices*	*Score*
13.	What are the advantages of using inorganic fertilizer ?	Providesdeficitnutrients-(1) Any other - (0)
14.	Is it necessary to eradicate excess aquatic weeds ?	Yes - (1) No - (0)
15.	Name some aquatic weeds	*Hydrilla, Eichornica, Typha*, etc. - (1) Any other - (0)
16.	Do you feel predatory and weed fishes are desirable in fish culture pond ?	No - (1) Yes - (0)
17.	Mention any two predatory and two weed fishes known to you.	Cat fish, murrel, Belchai, sosilu - (1) Any other - (0)
18	What in the manual method of eradication/ control of predatory and weed fishes ?	Repeated netting using small meshed drag net -(1) Any other - (0)
19.	Name any piscicide used in fish culture	Mohua oil cake/bleaching powder/Neem cake - (1 Any other - (0)
20.	Do you know the recommended dosage of mohua oil cake or bleaching powder ?	600-800 kg/ha of MOC - (1 Any other - (0)
21.	Name three Indian major carps	Catla, Rohu, Mrigal - (1) Any other - (0)
22.	Name three exotic carps	Common carp, Grass carp, Silver carp -(1) Any other - (0)
23.	Which is the fastest growing major carp and exotic carp ?	Catla and Silvercarp - (1) Any other - (0)
24.	Catla and silvercarp are surface feeders. Rohu is a column feeder Mrigal feeds on bottom vegetation Common carp in omnivorous	Yes - (1) No - (0)
25.	Which fish grows well in weed infested tanks ?	Grass carp - (1) Any other - (0)
26.	What is recommended rate stocking for irrigation tanks when CFC is practiced ?	4001-5000/ha - (1) Any other - (0)

Sl.No.	*Practices*	*Score*
27.	Do you know the recommended species combination for composite fish culture 3 SSP – 400 C : 300 R : 300 M (or 300 CC) 4 SSP – 300 C : 250 R : 150 M : 300 CC 6 SSP – 150 C : 250 R : 100 M : 200 SC : 100 GC : 200 CC	Yes - (1) No - (0)
28.	What is the ideal size of fish seed for stocking ?	30-40 mm - (1) Any other - (0)
29.	Whether supplementary feeding is necessary in CFC ?	Yes - (1) No - (0)
30.	Name the commonly used supplementary feeds.	GOC, Rice bran, silk worm, Pupae - (1) Any other - (0)
31.	What in the best method of feeding ?	Tray feeding/bag feeding - (1) Any other - (0)
32.	Generally supplementary feeding in provided at	2-3 per cent of body wt. - (1) Any other - (0)
33.	After stocking, once in how many days manuring should be done ?	Every month - (1) Any other - (0)
34.	Do you know the recommended manuring schedule to be practiced after stocking ?	200-250 kg/ha (RCD) - (1) Any other - (0)
35.	Do you known the indicators of oxygen depletion in fish pond ?	Gasping at surface/mass mortality - (1) Any other - (0)
36.	Is it necessary to stop manuring and feeding when pond water turns greenish ?	Yes - (1) No - (0)
37.	Name any fish disease that occur in fish culture ponds	UDS, Fin and tail rot, learnea - (1) Any other - (0)
38.	How do you control disease out breaks ?	By removing dead and infected fishes, by applying lime/salt/ash etc. - (1) Any other - (0)
39.	Is it necessary to check the growth after stocking ?	Yes - (1) No - (0)

Sl.No.	Practices	Score
40.	In general, after how many months of stocking the fish crop should be harvested ?	8-10 month - (1) Any other - (0)
41.	What should be the optimum size of harvesting ?	0.75 – 1.0 kg - (1) Any other - (0)

PART-III

Dependent Variables

A. A scale to measure attitude of fish farmers towards scientific fish culture

Sl.No.	Attitude Statements	Response categories				
		SA	A	U	D	SDA
I. Objectives and feasibility :						
1.	The composite fish culture is a viable economic activity					
2.	Composite fish culture encourages recycling of organic wastes like dungs, manures, etc.					
3.	Practice of composite fish culture involves negligible risk					
4.	Uncertainty of water availability in tanks due to unpredictive rainfall discourages the adoption of composite fish culture practice					
II. Technology transfer and development :						
5.	Adoption of composite fish culture increases employment opportunities for fishermen's community					
6.	Beneficiaries attend training programme under FFDA just for the sake of stipend.					
7.	Fisheries department is not equipped for adequate extension activities on composite fish culture.					
8.	Financial institutions/commercial banks do not readily advance loans to fish culture projects.					
9.	Generally standard (good size) fingerlings are not sold for stocking.					

Sl.No.	Attitude Statements	Response categories				
		SA	A	U	D	SDA
III. People's participation						
10.	Inspite of having a watch and ward, poaching of fish, cannot be prevented.					
11.	Use of poultry manure in village tanks is not an acceptable practice in the adoption of composite fish culture.					
12.	Fish pond excavation in swampy and low lying land for composite fish culture is a profitable venture.					
13.	It is difficult for a FFDA beneficiary to depend entirely on fish culture to make a living.					
14.	Fish farmers should adopt composite fish culture regardless of subsidy given by implementing agency.					
15.	Fish marketing lowers the social status of fish farmers in the society.					
IV. Production and risk factors :						
16.	Fish farmers should not discontinue fish farming if they sustain loss in the beginning					
17.	It is not possible for farmers to stock the required species of fish fingerlings.					
18.	To increase fish production, fish farmers should neglect villagers objection to manure the fish culture tanks.					
19.	It is difficult to obtain quality fish seed in desired proportion by fish farmers.					
20.	Incurring of losses due to diseases and pests is very low in tanks where composite fish culture is practiced.					

B. Adoption of recommended scientific fish culture practices

Sl.No.	Recommended Practices	Scores Assigned for			Reasons for Non or Partial Adoption
		Full Adoption	Partial Adoption	Non Adoption	
A. Pre stocking practices					
1.	Testing of soil and water				
2.	Application of lime for improving soil and water quality				
3	Control of aquatic weeds				
4.	Control of predators and undesirable fish fauna				
5.	Initial manuring with organic manures (RCD or poultry).				
6.	Application of inorganic fertilizer (urea and super phosphate)				
7.	Period of application of initial manures and fertilizer before stocking				
B. Stocking practices					
8.	Stocking of recommended number of different fish species				
9.	Density of stocking of fish seed (no./ha)				
10.	Ratio of stocking of selected species				
11.	Size of the fish seed stocked				
C. Post stocking practices					
12.	Use of supplementary feeds in prescribed proportion				
13.	Rate of feeding				
14.	Feeding method				
15.	Post stocking manuring with organic manures				
16.	Post stocking fertilization with inorganic fertilizers				
17.	Periodic netting to check the growth and health of fishes				

PART-IV

Constraints to Adoption of Scientific Fish Culture

Sl.No.	*Problems*	*Not so Serious*	*Serious*	*Very Serious*
A. Problems related to seed				
1.	Lack of knowledge about the qualitative seed			
2.	High cost of fish seed			
3.	Hatchery operators supply low quality seed			
4.	Non-availability of seed of different fish sp. at proper time			
5.	Any others (specify)			
B. Problems related to manure and fertilizers				
1.	Lack of adequate knowledge about manure and fertilizer (dose, time of application, methods etc.)			
2.	High cost of manure and fertilizer			
3.	Lack of manure and fertilizer			
4.	Lack of soil and water testing facility to the amount and type of fertilizer to be used			
5.	Any other (specify)			
C. Problems related to artificial feed				
1.	Lack of adequate knowledge about feed (dose, time of application and mode of application)			
2.	High cost of feed			
3.	Non-availability of high cost commercial feed			
4.	Any others (specify)			
D. Problems related to use of pesticides/insecticides/fish poison				
1.	Lack of adequate knowledge about the use of different pesticides, insecticides or fish poison			

Sl.No.	*Problems*	*Not so Serious*	*Serious*	*Very Serious*
2.	High cost of pesticides, insecticides and fish poison			
3.	Non-availability of desired pesticides, insecticides and fish poison			
4.	Lack of technical knowledge on judicious use of pesticides, insecticides and fish poison			
5.	Any others (specify)			
E. Problem related to social situation				
1.	Problem of poisoning in the fish pond			
2.	Problem of poaching and theft			
3.	Local non-cooperation of villagers			
4.	Problem of multiownership			
5.	Any other (specify)			
F. Problem related to marketing				
1.	Price fluctuation			
2.	Non-availability of assured market			
3.	Involvement of middle men is high			
4.	Any other (specify)			
G. Problems related to credit				
1.	Non-availability of credit facility in the locality			
2.	Complicated procedure for obtaining loan from the credit organization			
3.	Inconvenience due to distance to the credit organization			
4.	Untimely release of loan by the credit organization			
5.	Reluctant behaviour of bank personnel in advancing loan			
6.	Professional money lender charge			

Sl.No.	*Problems*	*Not so Serious*	*Serious*	*Very Serious*
7.	High interest rate on loan			
8.	Any other (specify)			
H. Problems related to transportation				
1.	Non-availability of self mode transport to bring produce to the market			
2.	Non-availability of Govt. or public mode of transport to bring produce to market			
3.	Harassment of police during transportation			
4.	Any other (specify)			

Date : *(Signature of the Investigator)*

Fish Production of Marine and Inland in India (1950-1951, 1960-1961, 1970-1971 and 1980-1981 to 2013-2014)

(In '000 Tonne)

Year	*Marine*	*Inland*	*Total*
1950-51	534	218	752
1955-56	596	243	839
1960-61	880	280	1160
1965-66	824	507	1331
1970-71	1086	670	1756
1973-74	1210	748	1958
1978-79	1490	816	2306
1979-80	1492	848	2340
1980-81	1555	887	2442
1981-82	1445	999	2444
1982-83	1427	940	2367
1983-84	1519	987	2506
1984-85	1698	1103	2801
1985-86	1716	1160	2876
1986-87	1713	1229	2942
1987-88	1658	1301	2959
1988-89	1817	1335	3152
1989-90	2275	1402	3677
1990-91	2300	1536	3836
1991-92	2447	1710	4157
1992-93	2576	1789	4365
1993-94	2649	1995	4644
1994-95	2692	2097	4789
1995-96	2707	2242	4949
1996-97	2967	2381	5348
1997-98	2950	2438	5388
1998-99	2696	2602	5298
1999-00	2852	2823	5675

Year	*Marine*	*Inland*	*Total*
2000-01	2811	2845	5656
2001-02	2830	3126	5956
2002-03	2990	3210	6200
2003-04	2941	3458	6399
2004-05	2779	3526	6305
2005-06	2816	3756	6572
2006-07	3024	3845	6869
2007-08	2920	4207	7127
2008-09	2978	4638	7616
2009-10	3104	4894	7998
2010-11	3250	4981	8231
2011-12	3372	5294	8666
2012-13	3321	5719	9040
2013-14	3439	6132	9572

Source : Central Water Commission, Govt. of India. (ON153), Ministry of Agriculture, Govt. of India. (14268), (ON463), (16361). Ministry of Statistics and Programme Implementation, Govt. of India. (ON821).

India Position in Aquaculture Production of Fish, Crustaceans, Molluscs, etc. in the World (2012)

(Value in 1000 US $, Quantity in Tonne)

Rank	*Country*	*Value*	*Quantity*
1	China	66212552	41108306
2	India	9248393	4209415
3	Vietnam	5807800	3085500
4	Indonesia	6715108	3067660
5	Bangladesh	3911495	1726066
6	Norway	5166850	1321119
7	Thailand	3316288	1233877
8	Chile	5993047	1071421
9	Egypt	2010814	1017738
10	Burma	1500569	885169
	World	**71518956**	**25524947**

Note : These countries are those with production of 140000 Tonnes or more in 2012.

Source : Food and Agriculture Organization (FAO).

Fish Seed Production in India
(1973-1974, 1978-1979, 1984-1985 to 2013-2014)

Year	*Fish Seed (In Million Fry)*
1973-74 (End of IV Plan)	409
1978-79 (End of V Plan)	912
1984-85 (End of VI Plan)	5639
VII th Plan	
1985-86	6322
1986-87	7601
1987-88	8608
1988-89	9325
1989-90	9691
Annual Plan	
1990-91	10332
1991-92	12203
VIII Plan	
1992-93	12500
1993-94	14239
1994-95	14544
1995-96	15007
1996-97	15852
IX Plan	
1997-98	15904
1998-99	15156
1999-00	16589
2000-01	15608
2001-02	15758
X Plan	
2002-03	16333
2003-04	19231
2004-05	20790.64
2005-06	21988.3

Year	*Fish Seed (In Million Fry)*
2006-07	23647.95
XI Plan	
2007-08	24143.57
2008-09	32177.21
2009-10	29313.17
2010-11	34110.83
2011-12	36566.43
XII Plan	
2012-13	34922
2013-14(P)	41450

Abbr. : P : Provisional.

Source : Ministry of Agriculture, Govt. of India. (16363)

Progress of Fisheries in West Bengal (1980-1981, 1990-1991, 1995-1996, 1998-1999 to 2005-2006 and 2008-2009 to 2012-2013)

Year	*Production of Fish (In '000 Tonne)*			*Production of Fishseed (in Million Numbers)*	*Production of Mechanisation of Boats (Number cumulative)*
	Inland	*Marine*	*Total*		
1980-81	340	30	370	2300	–
1990-91	555	125	680	7552	2100
1995-96	740	153	893	8180	2584
1998-99	824	171	995	8610	2974
1999-00	857	188	1045	8725	3762
2000-01	879	181	1060	8850	3622
2001-02	916	184	1100	9000	3856
2002-03	938	182	1120	9100	4175
2003-04	988	182	1170	11890	4481
2004-05	1035	180	1215	12200	4924
2005-06	1090	160	1250	12500	7994
2008-09	1295	189	1484	14000	9468
2009-10	1326	179	1505	12566	10606
2010-11	1246	197	1443	13453	11467
2011-12	1290.03	182.02	1472.05	13826	–
2012-13	1337.66	152.35	1490.02	15002	–

Abbr. : P : Provisional.

Source : Department of Animal Husbandry and Dairying, Ministry of Agriculture, and Ministry of Agriculture and National Institute of Rural Development and Ministry of Statistics and Programme Implementation.

Fishery Resources of West Bengal at a Glance

Particular	*Total Potential Resources (in Hactare)*	*Under Culture (in Lakh Hactare)*
Inland Sector		
Ponds/Tanks	276202	2.2
Beel and Baor	41782	0.21
Reservoir	16739	0.03
(a) River	172587	–
(b) Canal	80086	–
Sewage Fed Fishery	4083	0.04
Brackish Water Fishery	210100	0.39
Marine Sector		
Inshore Area(up to 10 Fathom depth)	770 km^2	770 km^2
Off Shore Area (10-40 Fathom)	1813 km^2	1813 km^2
Continental Shelf (up to 100 Fathom)	17049 km^2	17049 km^2
Coastline	158 km	158 km

Source : Economic Review 2001-2002, Government of West Bengal.

Fishery Water Resources in West Bengal and India (2014) (in Lakh ha)

States/ UTs	*Rivers and Canals (in Kms.)*	*Reservoirs*	*Tanks and Ponds*	*Floodplain Lakes and Derelict Water*	*Brackish Water*	*Total Water Bodies*
West Bengal	2526	0.17	2.76	0.42	2.10	5.45
India	195567	21.61	25.16	9.98	16.86	73.60

Abbr. : P : Provisional. Neg : Negligible.

Source : Ministry of Statistics and Programme Implementation, Govt. of India.

www.ingramcontent.com/pod-product-compliance
Ingram Content Group UK Ltd.
Pitfield, Milton Keynes, MK11 3LW, UK
UKHW021324190726
13845UKWH00003B/39

9 789389 605198